IN A NEW LIGHT

IN A NEW LIGHT
Histories of Women and Energy

Edited by

ABIGAIL HARRISON MOORE
AND R.W. SANDWELL

McGill-Queen's University Press
Montreal & Kingston • London • Chicago

ISBN 978-0-2280-0618-3 (cloth)
ISBN 978-0-2280-0619-0 (paper)
ISBN 978-0-2280-0756-2 (EPDF)
ISBN 978-0-2280-0757-9 (EPUB)

Legal deposit third quarter 2021
Bibliothèque nationale du Québec

Printed in Canada on acid-free paper that is 100% ancient forest free (100% post-consumer recycled), processed chlorine free

Funded by the Government of Canada   Financé par le gouvernement du Canada   Canada   Canada Council for the Arts   Conseil des arts du Canada

We acknowledge the support of the Canada Council for the Arts.

Nous remercions le Conseil des arts du Canada de son soutien.

Library and Archives Canada Cataloguing in Publication

Title: In a new light : histories of women and energy / edited by Abigail Harrison Moore and R.W. Sandwell.

Names: Harrison-Moore, Abigail, editor. | Sandwell, R. W. (Ruth Wells), 1955- editor.

Description: Includes bibliographical references and index.

Identifiers: Canadiana (print) 20210138505 | Canadiana (ebook) 20210145528 | ISBN 9780228006190 (softcover) | ISBN 9780228006183 (hardcover) | ISBN 9780228007562 (EPDF) | ISBN 9780228007579 EPUB)

Subjects: LCSH: Households – Energy consumption – History – 19th century. | LCSH: Households – Energy consumption – History – 20th century. | LCSH: Energy consumption – Social aspects – 19th century. | LCSH: Energy consumption – Social aspects – 20th century. | LCSH: Energy consumption – History – 19th century. | LCSH: Energy consumption – History – 20th century. | LCSH: Women – Social conditions – 19th century. | LCSH: Women – Social conditions – 20th century. | LCSH: Home economics – History – 19th century. | LCSH: Home economics – History – 20th century.

Classification: LCC HD9502.A2 I5 2021 | DDC 333.79 – dc23

This book was designed and typeset by Peggy & Co. Design in 10.5/13 Sabon.

Contents

Figures

IN A NEW LIGHT

Introduction

Abigail Harrison Moore and R.W. Sandwell

It matters what ideas we use to think other ideas ... It matters what stories we tell to tell other stories with ... It matters what stories make worlds, what worlds make stories.[1]

Energy is the food of life. Though typically taken for granted with its increasing invisibility in everyday life over the last century, energy has emerged as a discrete topic of discussion and analysis across all sectors of society in recent years. And no wonder. As the world confronts the realization that the fossil fuels that we rely on for almost everything – food, water, transportation, heating and cooling, medicine, communications, entertainment, and a whole world of goods – are resulting in catastrophic climate change, people are questioning en masse, and for the first time since the oil shock of the 1970s, the stubbornly held belief that higher energy consumption always means a better world. People are concerned that current very high levels of energy consumption in the so-called developed world cannot be sustained, let alone expanded to developing countries, potentially compromising people's comfort, convenience, and health. As governments and industries search for future solutions in new low-carbon strategies and technologies, historians are looking with renewed interest at the energy transition that first introduced the massive infusion of CO_2 into the atmosphere. Given the combination of fossil fuels' society-changing dominance around the world today and its catastrophic environmental consequences, the Industrial Revolution that ushered in its first massive use deserves, as J.R. McNeill recently opined, "to be elevated to the front of ranks among revolutions in the 250,000-year history of human society."[2]

As historians return their gaze to the Industrial Revolution with a new appreciation of changing fuel use as a historical agent, scholars within the emerging field of energy studies are acknowledging the importance of the historical context, and not just the technological and economic contexts, within which energy transitions have occurred. Earlier beliefs about the inevitability and direction of technological change have been tempered. A convergence amongst social, environmental, and energy historians has instead created what has been termed a new socio-ecological approach to energy transitions, expanding research beyond the new technologies of power and out to the contexts in which they have been used. A growing number of scholars have succeeded in putting "labor, human bodies and landscapes into the story of energy transitions."[3] They are re-evaluating long-held assumptions about the relationship between energy and capitalism, between prime movers and social formations, between fossil fuels and the environment, between "power to" and "power over."[4] Like others in the energy humanities, historians of energy continue to expand understandings of how persistent economic inequalities were worked out and worked through the corridors and carriers of power-as-energy. The roles of race, class, imperialism, slavery, settler colonialism, and environmental damage in energy transitions are being highlighted and their transnational, colonial, and postcolonial contexts explored.[5]

But in surveying with a critical eye the rich and growing literature embraced within the social history of energy, there is a notable absence: women. Arguably no aspect of energy's history is less developed than gender, and no topic less explored than women's relationship to the last great society-changing transition to fossil fuels. Studies of energy still focus overwhelmingly on the technical systems of energy supply that define the industrial world. As a result, we now have a rich and growing analysis of men's inventions, men's labour, and men's planning and development of systems for financing, organizing, selling, running, repairing, and maintaining the new networks of power, including those of coal, oil, gas, and electricity. Historians have documented the extensive environmental and social damage that has followed in their wake.[6] There are some important studies of gender and masculinity in energy history.[7] We also have some very good analyses of the wider national and international political, cultural, and economic implications of energy transitions, augmenting our understanding of how these were created, mediated, and experienced. These are undeniably important topics, and ones that we would expect to intersect with women's lives, and in a variety of ways. But, with a few important exceptions, these intersections have not been explored.

Women do not appear very often in histories of energy, but when they do, their relation to energy seems very different than men's. As chapter 1 explores in more detail, this is because energy history has largely been defined as that of the industrial forms of energy, and particularly by developments within the new network systems of electricity, oil, gas, and nuclear power. The work of Graeme Gooday, which places Euro-American women at the centre of various narratives of energy history, has been an influential exception. Other historians have explored women's roles as energy actors, but such work has generally been classified as women's or cultural history, or the history of the family, and is not generally seen as energy history, nor is it read by energy historians. A number of welcome studies, for example, have looked at the professional women who serviced the new generation of home energy consumers, both the corporate "demonstrators" of new appliances, and the growing band of professional domestic science and home economics educators.[8] These studies are important in documenting changing patterns of women's labour, and women's increasing political consciousness about the nature and status of their domestic roles.[9] The majority of these studies has focused on the material culture of the twentieth century, often exploring energy transitions through the development of national grids and infrastructures that assumed energy choices to be about gas and electricity, reading this as a moment of men's leadership in transforming technology and invention.

This volume argues that women's near invisibility in energy history is a problem, and one that requires explanation and redress. No identifiable group likes to be left out of history.[10] But the omission of women from energy history is particularly problematic. While abundant historical evidence confirms that women neither initiated nor responded to the transition to fossil fuels in a coherent and essentialized manner across the intersectionalities of race, class, and gender, gender nevertheless appears over and over again as a defining feature in the myriad changes associated with industrialization and modernity. The essays in this volume focus on the early-industrializing societies of north-western Europe and North America, and on the nineteenth and twentieth centuries, for which the near total absence of women in the early history of the new centralized power systems is a case in point: if gender is *not* a relevant category of analysis for understanding their development, why were women so consistently absent in those early years? Moving from history to historiography, and from absence to presence, abundant evidence suggests that meaningful generalizations can be made about the ways gender influenced people's experience of the transition to fossil fuels and how gender worked to structure roles within industrializing societies. If, as we will argue, women's experiences of and roles within the transition to fossil fuels have been different from men's – if gender has been a factor in organizing and structuring energy transitions in these early-industrializing societies – perhaps we can learn something about the history of energy by identifying and exploring how and why gender steered industrial societies' encounters with energy.

The chapters do more than add these lost voices and perspectives to energy history. As Frank Trentmann phrased it, while energy historians have focused so intently on the supply side of the new networks of power, "when we instead turn our eye to what energy is used for, different actors are allowed onto the stage: housewives, children, the elderly, lighting designers and appliance salesmen and women, as well as office workers relying on energy-hungry equipment at their workplace."[11] We argue that opening up the stage to new actors playing new roles also changes the plot. Including women obliges historians to look in new ways at possibly unfamiliar places to understand the changing role of energy in society. This volume unapologetically returns the historians' gaze to women's domestic lives as a way of deepening our understanding of gender in the history of energy, and energy in the history of women and gender.

In a New Light

Energy history is indeed changing the ways that historians understand the past, but the absence of women continues to both reflect and propagate some misapprehensions about the complex relationships between energy and society. The chapters in this volume are organized in generally chronological order. Some of the essays demonstrate that many women embraced narratives of women's emancipation, and became scientists, professionals, politicians, and protesters within new energy frameworks. Other essays explore the rich diversity of women's domestic work as wives and mothers. As the historians in this collection argue, while women typically experienced the transition to modern energy in different ways to men, and generally from the vantage point (though not always from the physical location) of home and family, those ways varied across time and space, and included protest, experimentation, and resistance, as well as accommodation. And their experiences, en masse, had some unpredictable outcomes in terms of women's changing role within society. The essays in this collection help to explain how and why women's lives, and their agency, were manifested through their experiences of the transition to modern fuels.

While individual essays explore particular case studies and issues in women's transition to modern energy, the volume as a whole explores a rich variety of ways in which gender and energy have intersected in women's lives in some distinct gender-specific ways. The energy transitions that drove the Industrial Revolution produced a fundamental change in the way that men and women's lives were organized and structured. These chapters map out this history by focusing on the nineteenth and twentieth centuries. Britain, as the first country to industrialize, plays a significant role in the volume that follows, while chapters on Germany, Ireland, and Canada explore similarities and differences in the process of industrialization geographically and over time. While our authors write from a variety of approaches and geographical contexts and use different research methods to explore different times, the chapters in this book aim to draw out the cross-cutting themes that help us to see women's vital and active roles in energy transitions in a new light. Most of the essays in this collection explore women's varied responses to this gender-specific loss of economic, social, and political power that came with industrialization. What follows are a series of carefully drawn case studies of women's experiences of and agency in the history of energy in Europe and North America in the nineteenth and twentieth centuries.

Chapter 1 provides something of a backstory to both women's distinctive role in the energy transition known as the Industrial Revolution, and to their historiographical invisibility within energy history. It sets out to explain how the field of energy history emerged without including gender or the home as significant categories of description or analysis, and what is to be gained by including them. Energy history has long tended to focus on the invention and production of fossil fuels and electricity, and their distribution through the market activities that have so preoccupied economic historians. Men were the energy agents in the narrative. But the silence of energy historians when it comes to women is nevertheless puzzling in light of the detailed research by an earlier generation of social historians who clearly identified the family, the household, and then gender as categories of analysis every bit as integral to the changes associated with industrialization as the newly emerging class differences.[12] The chapter explores studies by a phalanx of historians in the 1970s and 1980s of social, demographic, family, and women's history and historians of masculinity from the 1990s who established a broad consensus that industrialization remade gender distinctions, and in ways that generally disadvantaged women. The roles of men and women began to diverge sharply as households became more dependent on wages, and wages became increasingly identified with male breadwinners.[13] A nascent male labour movement and a growing bourgeois ideology increasingly consigned women to a domestic role inside the home. At the same time, legal and ideological prescriptions constrained women's access to waged work outside the home and relegated them, ideologically if not always materially, to unpaid domestic labour. As women's access to paid work was constrained by conventions and protective legislation, their unpaid labour in the informal economy also became increasingly disparaged, undervalued, and indeed was not recognized as work.[14] Numerous studies over the years have demonstrated that women's unpaid labour, while largely invisible, remained, in fact, vital in supporting households and community through such activities as cooking, gardening, scavenging, piece work, laundry, borrowing, sharing, and of course the caring and social reproductive work for which women continued to be primarily responsible.[15] The chapter concludes that while the ideology of separate spheres has been tacitly invoked to somehow naturalize and explain women's distinctive invisibility in energy transitions, it is instead an unacknowledged *result* of industrialization. Women's historiographical invisibility not only obscures the importance of gender inequality in the history of energy transitions but, in the process, recapitulates the ongoing gender hierarchies, the different *valuations* of men's and women's labour, that provide structural, if largely invisible, support to the industrializing world. What that looked like – how energy was experienced by women and was in ways structured by their work and their leisure within the home – is the subject of subsequent chapters.

In chapter 2, Karen Sayer explores the material cultures of lighting, opening up a new approach to thinking about the experience of our oldest form of lighting technology – the candle. Sayer documents the important role that women played in energy in the pre-industrial era, confirming the time, skills, and expertise required in providing and enhancing candlelight, and the gendered and class dimensions revealed by the English archives that document the innovations in candles to create brighter, better-burning, and smokeless light. Chapter 3 turns to Canada. Ruth Sandwell's closely drawn analysis of women's experiences of energy in daily life within their domestic roles explores a seldom-examined aspect of women's transition to the new fuels: their fear and anxiety about using new and potentially dangerous fuels. It also highlights the ways in which women actively adjusted their behaviours in the home to accommodate such problematic energy transitions.

Picking up on these themes of anxiety and action, Abigail Harrison Moore looks in chapter 4 at women's new roles in the late nineteenth century, as both consumers in the new wealthy homes of the industrial era, and as new design professionals hired to advise and reassure women as they integrated new kinds of energy into the home. She shows how women reshaped the contours of the "non-productive" household sphere to which industrialization had relegated (respectable) women, finding new and more powerful roles as consumers and as paid professionals within the confines of gender-specific respectability.

As many of the chapters in this volume touch on directly and indirectly, the growth of the relatively wealthy middle class, with a new and special kind of home, was a crucial factor in energy transitions – transitions with which women were intimately involved. Women, and particularly the growing number of middle-class women, were pressured to see their role in the home as a "moral" one, creating an unpaid and unrecognized duty to provide a healthy and beautiful home for their working husbands to return to. As England underwent rapid social, cultural, and physical change as a result of the Industrial Revolution, and the cities grew, commerce and industry began to be seen as "increasingly brutal and deceitful ... To the middle classes in the nineteenth century, the home stood for feeling, for sincerity, honesty, truth and love."[16] This led to the requirement for and rapid development of advice literature, in which "ideas about morality, religion, political economy, gender, social status, taste and comfort were all played out."[17] The gendered expectations in this period can be seen clearly in advice given in a chapter on "The Management of the Home" published in 1867: "It is not ... merely in his wife that a husband should look for and expect beauty within the walls of his own dwelling. It should, as far as possible, be made to appear in everything with which he is surrounded ... the influence of such deposition of things in soothing the appetites with which the outward life of man is assailed."[18] With the concurrent rise of capitalist structures for consumption, illustrated by the growth of the department store, women were given "choice" – but with choice came inevitable anxiety about "getting it wrong," as well as the pressure to create a house that reflected their morality as well as their taste, framed by the idea that they were *limited* to such activity. As John Ruskin stated in *Sesame and Lilies* (1865) "The woman's power is not for rule, not for battle – and her intellect is not for invention or creation, but for sweet ordering, management and decision."

The pressure on the middle-class homemaker generated some of the first professional opportunities for women from the same class, as advisors and designers, work that they used to engage their patrons in the call for suffrage. In 1888, an article in the English *Art Journal* commented that "We require work, not so much for poorer and less highly educated women, who naturally take to the avocations where physical strength and endurance are most required, as for the multitude of middle-class women who have to struggle for a living. We require also, for the proper doing of that work, a more systematic and thorough education than that

which women of the middle class generation receive ... The housemaid,
who has been taught how to clean a room thoroughly, has received a
more serviceable education than the average young lady." The aim of
the Women's Arts and Industries' section of the Glasgow International
Exhibition that this article referenced, therefore, was "to endeavour to
extend the knowledge of what women can do," and as such the decoration of the home, including the decision as to how to heat and light
the home, became an area in which women began to fight for a right
to be professionals, and to access the requisite training.[19]

It is interesting that women's first professional engagement in energy
transitions was via craft, design, and decoration. A number of feminist
art historians have argued that the way in which craft was seen as a
"lesser" art form enabled women to find a space in which to break
through gender boundaries. Although women were active within the
Arts and Crafts movement, as makers and writers, they were unable to
train as architects until the 1890s, given that before the Married Women's
Property Act was passed, they were unable to make contracts or be sued
in their own right and therefore could not assume many of the professional responsibilities of the architect.[20] In terms of the development of
professional opportunities for women in the engineering sphere, these
were likewise often hidden behind titles like "wife" and "Mrs," as illustrated by the case study of Mrs J.E.H. Gordon, and her 1891 publication
Decorative Electricity. In this book, the electrical engineering work was
articulated as the province of James Edward Henry Gordon, with his
wife, Alice, relegated to making the decorative decisions, whereas, as
Graeme Gooday has uncovered, Alice's abilities as an engineer were in
many ways equal to those of her husband.[21] The very fact that she was
forced to publish under her married name, using her husband's initials,
obscured Alice's identity as an electrical expert in her own right.

Turning to the objects created to help sell and justify the installation
of gas and electricity in the home, Ruth Schwartz Cowan's pathbreaking
work *More Work for Mother*, published in 1983, was the first to clearly
articulate the problematic nature of equating modern technologies
with liberation, arguing that women had more work to do, as new
appliances raised the standards of, and reduced men's involvement in, the
seemingly unending tasks of homemaking.[22] Ever since, historians have
remained skeptical of the emancipatory potential of new labour-saving
appliances in ensuring women's increased leisure, let alone their role
in promoting equality. A number of the authors included here take
Schwartz Cowan's work and expand on it through their more explicit
focus on energy, finding ways to integrate women's energy work more
firmly into the larger cultural, political, and economic complexities and

narratives of the nineteenth and twentieth centuries, rather than to see it as an exception, or as existing outside of them. Furthermore, while we aim to re-examine, in a much more active way than has previously been done, the energy responsibilities derived from women's roles as servants, wives, and mothers in the home, we also harness a much more open category of "woman," to focus on women's agency as manifested through their varied experiences of the transition to fossil fuels.

In the early twentieth century, as the example that Graeme Gooday explores in chapter 5 demonstrates, women, with the expertise they won through the opportunities opened up by the First World War, managed to secure at least some power within the engineering world. Men's lack of authority over women in domestic energy consumption helped allow for the creation of engineering associations for a new elite of middle-class women, to help them become indispensable players in the enhancing demand for the electrical supply industry. A broad-ranging coalition of women from all social classes acted as agents to undermine the male monopoly on engineering and technical education while reconfiguring, but not overturning, the pervasive belief that women were responsible for domestic energy management. This burden of responsibility continued even as the role of spousal support for engineers disappeared with the rise of corporatized engineering culture, which saw the earliest association dissolved as it reached the limits of its transformative capacities later in the twentieth century. Gooday's chapter contrasts with Schwartz Cowan's view of women at this time as powerless to shape the new electrical energy supply industry, as well as with Thomas P. Hughes's contemporaneous *Networks of Power* and its historiographic portrayal of an industry exclusively managed by men, without reference to women's expertise, needs, or interests.[23] Gooday develops this theme of women's agency by examining the different ways that they were able to embrace the new opportunities offered by the industrial regime by adopting the new profession of electrical engineer, actively challenging gender stereotypes, and finding power though association. He shows how British women reshaped and expanded on existing gender roles to become new kinds of professionals, devoted to helping women-as-consumers negotiate energy transitions in their homes.

In chapter 6, Sorcha O'Brien looks at women's extensive energy work within the domestic sphere in Ireland, and the collective organizations that brought women together. She explores the pride and the frustrations that women felt in the mid- to late twentieth century as a result of using their new skills and labour to provide energy in the home, and their hopes and fears for the new forms of energy. Women working together to bring about change is also at the heart of Petra Dolata's chapter on women in West Germany. In it, she explores more explicitly the ways that women managed some of the contradictions, dangers, and inconveniences imposed on them by industrial energy. She focuses on housewives in a region in Germany where the material conditions of coal energy production – the dangers of its production, and the filth and pollution generated by its consumption – had not only long imposed gender divisions of labour, but also imposed additional burdens on women's work within the home, as well as on the health and well-being of their families.

In the final chapter of the volume, Vanessa Taylor closes the book with her essay "Anthropocene Women: Energy, Agency, and the Home in Twentieth-Century Britain." Both a microhistory and reflection on women, energy, and the home, the essay skilfully brings together the material conditions of energy production, the cultural nuances of energy in daily life within the home, and the professional opportunities opened up for women in Scotland in the mid-twentieth century. It provides a nuanced description and analysis of the factors influencing women's role in the transition to modern energy. As she concludes, in the active discussion and politics of the current period of energy transition, "now seems a good moment to highlight once more women's agency in the transition to energy-intensive domestic technologies of the past centur[ies]."

Notwithstanding the differences in the various case studies on women and energy presented here, certain themes run through this collection. Most of the essays show women adapting to new forms of energy, albeit while also exhibiting certain anxieties and ambivalence in the process. Throughout these essays, we can also see women *enhancing* energy, creating the demand and the design for new forms of energy in the home. Women *enfranchised* energy, both by actively seeking out energy transitions, and in other ways by informing, and being informed by, wider ambitions for equality. These themes highlight women's pivotal roles in the consumption and production of energy in the home. The essays also explore women's engagement with and responses to particular energy transitions, their changing energy-related identities, and their behaviours and experiences. We can see as well women's politically informed associations, their self-identified collectivities, and their gender-specific normativity – and occasionally their self-styled exceptionalism – in energy-related behaviours. We see women as political energy-activists and glimpse some of the factors influencing women's energy decisions. The authors in this collection show us the role of energy objects, aesthetics, and materiality, and women's engagement with time and teleology through their energy practices. Finally, these chapters document not only women's agency and successes, but their culpability and their victimization through energy use, and within the new structures that defined power in the industrial world.

Our ambition for the volume, and in our individual work, has been to empower women in these histories rather than simply reinsert them. This is not just about women's experiences of energy transitions, but also about reframing the power relations in these histories of important moments of change. We want to move away from a deficit model of women and technology in the home, which, when women are mentioned at all, has often framed them as passively or grudgingly accepting of the decisions made by men. Instead, we propose that, through engagement, adoption, adaptation, consumption, and protest, women took on positive, influential, and active roles in energy decisions. By giving value to the histories of this work, we want to collectively present a major challenge to the ways in which energy history is currently undertaken and to suggest that the lessons learned from observing women's roles in the great energy transitions of the Industrial Revolution may help us think about the role we will take on in the urgent move to a carbon-neutral future.

Notes

1 Haraway, "SF: Science Fiction."
2 McNeill, "Cheap Energy and Ecological Teleconnections," 30.
3 Barca, "Energy, Property and the Industrial Revolution," 1314. For more about the changing direction of energy history, see also Saelens, "Review of *The Path to Sustained Growth.*"
4 See for example, Huber and Jones, *Routes of Power*; Malm, *Fossil Capital*; Johnson, *Carbon Nation*; McNeill and Engelke, *The Great Acceleration*; White, *The Organic Machine*; Barrett and Worden, eds., *Oil Culture*; Huber, *Lifeblood*; and Szeman, et al., eds., *Fueling Culture: 101 Words for Energy and Environment*. The new socio-ecological approach to energy has stimulated extensive and detailed empirical and comparative histories of changing energy consumption and production, set in their social and cultural context. Data-heavy research has been undertaken by Kander, et al., *Power to the People*; Fouquet, *Heat, Light and Energy*; and Unger and Thistle, *Energy Consumption in Canada*.
5 Energy humanities is a rich and rapidly growing field of study. See for example, Pomeranz, *The Great Divergence*; Ghosh, *The Great Derangement*; Mitchell, *Carbon Democracy*; Agarwal, "Does Women's Proportional Strength

Affect Their Participation," 98–112; Massell, *Quebec Hydropolitics*; Waldram, *As Long as the Rivers Run*; and Rubio, "Will Small Energy Consumers Be Faster in Transition?," 50–61.

6   Environmental historians have provided an impressive number of studies documenting the environmental damage caused by extracting, transporting, and processing energy, including that from uranium, oil and gas, and their byproducts, particularly insecticides and fertilizers, as well as studies documenting the growing spectre of climate change. Well-known examples include Carson, *Silent Spring*; Zalen, *American Lucifers*; Melosi, *Coping with Abundance*; and Black, *Petrolia*. A number of contemporary studies have focused on the effect of pollution, mostly from oil and gas combustion and from petrochemical products, on women's bodies. See for example, Murphy, *Sick Building Syndrome and the Problem of Uncertainty*.

7   See for example, Holmes, "Making Masculinity," 39–48.

8   See for example, Clendinning, *Demons of Domesticity*; Goldstein, *Creating Consumers*; Harris, "The South Carolina Home in Black and White," 477–501; Jellison, *Entitled to Power*; and Stage and Vincenti, eds., *Rethinking Home Economics*.

9   See for example, Horowitz and Mohun, *His and Hers*; and Forty, *Objects of Desire*.

10 There has been, as feminist historians are well aware, considerable debate about whether women are, indeed, an identifiable category. This volume assumes that while "women" or "men" are not comprised of an essential, unchanging identity as a result or cause of their biology, gender is a category of analysis that has played a significant, through varied and changeable, role throughout history. Not discounting that gender identity, both ascribed and embraced, exists within a complex intersectionality of race, class, and various other signifiers, chapter 1 argues that the transition to modern energies in nineteenth- and twentieth-century North America and Britain was brought about in the context of a sharpening of gender distinctions and inequalities, most particularly manifested in ideologies of the "separate spheres," and the legal, political, economic, environmental, and cultural changes that this brought about for those gendered female during this period.

11 Trentmann, "Getting to Grips with Energy," 8.

12 See for example, Scott and Tilly, *Women, Work and Family*; Kriedte, Medick, and Schlumbohm, *Industrialization before Industrialization*; and Wrigley, *Continuity, Chance and Change*.

13 Landes, *Women and the Public Sphere in the Age of the French Revolution*; Berg, "Women's Work and Mechanization"; Lehning, *The Peasants of Marlhes* and *Peasant and French*; Gullickson, *Spinners and Weavers of Auffay*; Accampo, *Industrialization, Family Life and Class Relations*; and Griffin, *Breadwinner*.

14 The ideological foundation of this revaluing of women's work is explored in Valenze, *The First Industrial Woman*, in the British case, and in the American case by Boydston, *Home and Work*; Folbre, "The Unproductive Housewife"; and Nicholson, *Gender and History*.

15 Boydston, "To Earn Her Daily Bread," 13–14; Pennington and Westover, *A Hidden Workforce*; and Newton, *The Feminist Challenge to the Canadian Left*.

16 Forty, *Objects of Desire*, 101.

17 Kelley, "Housekeeping: Shine, Polish, Gloss and Glaze," 94.

18 Sherer, *The Family Friend*, 201–3.

19 "Glasgow International Exhibition," 6.

20 See for example, Callen, "Sexual Division of Labour in the Arts and Crafts Movement"; Walker, "Women Architects"; and Parker and Pollock, *The Old Mistresses*.

21 See Gooday, *Domesticating Electricity*.

22 Schwartz Cowan, *More Work for Mother*.

23 Hughes, *Networks of Power*.

1

Changing the Plot:
Including Women in Energy History
(and Explaining Why They Were Missing)

R.W. Sandwell

Introduction

Where are the women in energy history? Women are almost entirely absent from general histories of energy and energy transitions, just as energy as a discrete and identifiable phenomenon or force is largely missing from histories focusing on women and gender. The home is not the only place in which to explore women's experiences of energy transitions, nor is it the only place in which to observe the role of gender in structuring the society-wide changes ushered in by the transition to fossil fuels. But for studies like this one exploring energy use in nineteenth- and twentieth-century Western Europe and North America, it is an obvious location – ideological, aspirational, and material – to find women, and to clearly identify them as actors in the history of energy.[1]

In one sense, there is little new here. Although climate change has only recently redirected historians' attention back to the Industrial Revolution to re-evaluate it as the starting point of massive fossil fuel burning,[2] historians have long argued the importance of the household in establishing the particular conditions for economic growth associated with early industrialization.[3] They have shown how women, families, and households were profoundly intertwined with the social, economic, political, and cultural changes that marked the long Euro-American transition from a rural and land-based society to an urban, capitalist, and industrial one. The household itself changed significantly, and with it, women's economic roles and their status. Industrial-era households maintained some pre-industrial features, particularly their nuclear structure: while households often included paying boarders and paid servants, they continued to be largely kinship-based and hierarchically organized by age and gender. Labour within industrial-era households continued to include a range of activities remunerated "in kind" as well as "in cash." But it was only with industrialization that so-called productive labour moved out of the home and into workshops and factories. Gender divisions sharpened. Adult men's "breadwinner" wages became an increasingly important component of household support in urban, capitalist society. The industrial-era home gradually became identified as a place where housework and care work were carried out almost exclusively by women and girls as unwaged wives, mothers, and daughters, and as badly paid servants. As a generation of women's and labour historians have argued, women have continued to provide vital support to their families (and the economy) through a range of labour, including subsistence, care work, housework, petty commodity production, and their waged work outside and inside their family home and in the homes of others.

This chapter builds on historians' studies of Euro-American women's extensive and varied activities within the domestic spaces of the home in the industrializing era. It argues that the new, increasingly domestic roles women were allocated in industrial society were, like men's increasing reliance on waged labour outside the home, a result of the transition to fossil fuels: together they comprised the new gendered economic re-arrangements ushered in by industrial capitalism. While the household was no longer the centre of economic production, Victorian and Edwardian Euro-American families continued to play "a key role in the earning and allocation of resources." Well into the twentieth century in industrialized countries, "The family, not the state, emphatically remained the institution responsible for the distribution of wealth."[4] Historians have acknowledged the importance of the waged work of women as well as men in industrializing societies; this chapter focuses on the household labour of women and girls, arguing that while undervalued, underpaid, and too often overlooked then and now, domestic work made vital contributions to industrial society's energy metabolism, providing important energetic foundations for industrializing capitalist society.

The home has clear advantages for those wanting to identify and understand Western women's experiences and roles in nineteenth- and twentieth-century energy history. But it also presents some vexing conceptual problems for historians. Defining society as being comprised of gendered separate spheres of private homes and public places outside of them has the advantage of summing up a particularly powerful ideology emerging by the early nineteenth century, one in which newly defined gender roles sharply defined how and where Euro-American people should live, love, and work. But historians have long documented differences between prescription and description in this era: clear evidence shows women working for wages, and engaging with the public, political, and economic world in a variety of capacities throughout the nineteenth and twentieth centuries. Feminist historians have taken the critique of separate spheres further, showing how its underlying assumptions continue to influence the writing of history. As Leonore Davidoff and Catherine Hall described it, "The world of production and the state have been systematically privileged as central to historical understanding"; they accused scholars of inheriting a "double view" of the social order and consigning women to the home and family, sites which are accorded no conceptual or analytic importance in social theories.[5] As Linda Nicholson put it, the problem thererfore is not simply that historically "the divisions between these spheres are not as rigid as we are led to believe," but also that "conceiving them in such a manner obscures the realities of women's lives."[6] In other words, industrialization did create new roles for men and women, and women's lives generally did differ substantially from those of men, but the concept of separate spheres does not give us good enough tools to explain the gendered inequality of industrial societies, or to understand the contours of women's lives.[7]

Women's energy history is a case in point, for arguably the ideology identifying the home as a separate sphere continues to *naturalize and render opaque* women's distinctive relationships with both the economy and energy in the industrializing era. Though feminists have pointed out that the unpaid housework and care work of women has been "necessary and time-consuming, involves skill and physical exertion, and is essential for the functioning of the market economy,"[8] and notwithstanding the fact that a generation of historians of "women and masculinity have succeeded in establishing domestic life as a subject to be taken seriously," the home seldom appears in mainstream histories of any kind.[9] If it does, it is identified as being of social and personal rather than economic and political significance, and thereby trivialized and ignored. As a result, after more than a hundred years of sociological and historical research, it is still the case that "the most serious gap in our knowledge relates to the nature of housework and care work and how this work changed over time."[10]

If women's encounters with energy centred around the household, and if the household is largely excluded from definitions of both labour and the economy, it is little wonder that women have been absent from energy history. A chorus of feminist historians have argued that "making sense of women's lives therefore requires us to move into unfamiliar terrain."[11] Women were *not* invisible as energy actors, but their increasingly gender-specific role assignment within industrial societies often makes them difficult to see. This chapter describes how the field of energy history emerged without gender or the home as significant categories of description or analysis, and what is to be gained by including them.

Energy Use before Industrialization

To better understand the history of women, gender, and energy, a few preliminary definitions, descriptions, and historiographical discussions are in order. Energy can be defined as the ability to do work, and power as the rate at which it is done. Heat is generally recognized as a special kind of energy, one that performs the equivalent of work through temperature-determined chemical changes.[12] Early in its history, the human race was unique in learning to generate its own energy by controlling fire for a variety of energetic tasks: artificial lighting, space heating, cooking and otherwise preserving food, and for smelting and forging a wide range of materials.[13] The food and fuel consumed by people was provided by the sun, via photosynthesis. People ate animals that likewise relied on plants, and also relied on both to provide fuel oils for lighting, cooking, and heating. Animals provided energy for transportation, and stationary and motive power to supplement human muscle power. Reins, yokes, "horse powers," and "sweeps" facilitated the transfer of energy from muscles to mechanical power for a variety of ploughing, cutting, grinding, twisting, pressing, and milling tasks.[14] People also transformed a variety of animal and vegetable products into clothing and shelters that allowed them to better conserve heat and food energy. Finally, the sun also creates the complex hydrological systems that provide wind and water, also vital supports for life on earth. Over millennia, people have harnessed energy from both, developing techniques and machines, such as water wheels, turbines, and windmills, to transform their kinetic energy to mechanical energy to be used as a supplement to muscle power. In sum, before people began to burn fossil fuels to meet their energy needs, the materials on which people relied to make a living and a life – their tools, their housing, their transportation, their energy for food, heat, and power – were derived directly from the land and nearby waters, and directly and indirectly from the sun.

But if the sun provides the energetic foundation of life, until quite recently it was muscle power that acted as the main intermediary between the various kinds of sun-derived energy and people's access to it. As the first law of thermodynamics dictates, energy is never really lost, or spent, or consumed – it is only converted from one form to another. Throughout most of human history, muscle power has been the main energy converter on which people relied. Wind, water, and wood of course provided energy, but only after inventions and interventions provided by the muscle power of men and women, and often their children. In the endosomatic (human-powered) world of organic energy, human muscles provided more than 70 per cent of mechanical energy.[15] The muscles that mediated between people and the environments from which they were converting energy belonged to both men and women.

Indeed, it is no exaggeration to say that most men and women spent most of their time throughout human history in personally and viscerally wresting energy from the environment, and converting it in various ways to support life. Planting seeds and harvesting edible plants, tending and slaughtering animals, cutting down and otherwise processing trees for home heating and cooking, processing and transporting goods to market by horse and cart, filling oil lamps, cooking and eating dinner, or nursing an infant – all are ways in which people converted energy into a usable form by means of human and sometimes animal muscle power, along with their skills, knowledge, and expertise. A recent discussion of energy and everyday life summed up the relationship between people and energy this way:

> energetic flows may be seen as being so fundamental to the
> work of life, society and community that they constitute a
> "social metabolism" – a telling reversal of the metaphor of the
> "organic machine." Understood in these terms, any given society
> can be framed as an open system of energy flows that must be
> maintained for it to function.[16]

The energetic work of pre-industrial peoples occupied so much of their time, and comprised so many of their daily activities, that the relationships between people, energy, and environment are plainly visible for the historian to see in the practices of everyday life. By contrast, as energy came increasingly from underground sources, and was transmitted to consumers as products and services delivered through a series of increasingly automated literal and figurative black boxes, the relations between people, energy, and the environment became increasing difficult to perceive.[17]

Energy Historiography: Defining the Main Story of Energy Transitions

Historiographical trends in energy history did not determine women's energetic contributions, but understanding these trends can help to explain women's current near-invisibility in energy history. The history of human society could, indeed, be framed as an open system of energy flows, as a kind of social metabolism that includes men, women, and households, but it seldom has been.[18] Energy is undoubtedly the foundation of life on the planet, but if human history has been tightly interwoven with, and in many ways bounded by, the ways in which people have converted energy from the environment for their own uses, these relationships have not been the focus of most energy history. When the role of energy in human history has been discussed at all, it is a history much more narrowly defined as the transitions in energy production and use associated with the Industrial Revolution. Accounts have typically been dominated by triumphalist stories of how "man" harnessed new kinds of power over nature through his clever discoveries and inventions of electricity, fossil fuels, and later nuclear power. This way of thinking about energy and society has a long history, and it is having a long legacy.[19]

From the eighteenth century, early scientific experiments with coal, coal gas, and electricity inspired the popular imagination with new ideas, utopian and dystopian, about the kinds of human progress and prosperity that new forms of energy use could bring.[20] In 1884, Arnold Toynbee identified steam power and the division of labour as the key characteristics that changed society during the Industrial Revolution. In 1907, German philosopher and scientist Wilhelm Ostwald developed a theory of energetics, arguing that "the history of civilization" is "the history of man's advancing control over energy."[21] By the mid-twentieth century, some social scientist were placing changing energy use at the centre of historical change. What became known as the "energy stage" theory posited that it was "man's" ability to harness energy "in excess of that expended to make energy available" that characterized modern industrial societies, and which gave them their economic, social, and cultural superiority.[22] They contrasted high-energy-consuming industrial societies with "backward" low-energy, agricultural ones. As Carlo Cipolla summarized in 1962, "the more successfully man controls and puts to use non-human energy, the more he acquires control over his environment and achieves goals other than those strictly related to animal existence."[23]

Energy-stage theory not only recapitulated hierarchies of race in this narrative of civilization, but, as Carolyn Merchant and others have argued, it also rested on a profoundly gendered cosmology and ecology wherein science and capital are combined into a new kind of dominant masculinity, defined as mechanistic, rationalized, and competitive, and in opposition to the natural, nurturing, emotional, and non-economic world of the feminine.[24] Discussions about whether "woman" is a coherent category of analysis have persisted. Notwithstanding accusations that environmental history has marginalized women, feminist historians have taken up differing positions on whether or not women have a unique relationship to the environment via their particular role as mothers and their association with the "nurturing home." Debates are still circling around whether the observable historical differences between men and women's relationship to the environment, and to environmentalism, are the result of biological destiny or historical contingency.[25]

There was a flurry of interest in energy, some of it historical, during and shortly after the oil shock of the 1970s, and a number of historians, economic and otherwise, challenged the narrative of progress and development.[26] These quickly disappeared from sight, particularly within the North American literature, in the 1980s. Through the later twentieth century, general discussions about the nature and role of energy in society waned, along with stage theory and a general interest in the Industrial Revolution. Historians working in the subfields of environmental history were important exceptions. Gaining momentum in the twenty-first century, environmental historians have detailed the catastrophic consequences of fossil fuel use for people's bodies and for their environments, and argued that this damage must be included in descriptions of and valuing, financial and otherwise, of the economic growth ushered in by fossil fuels.[27] They have been among those arguing that the environmental damage, like the social disruptions and increasing inequalities that so often accompany the transition to modern energy, is not an "externality" of the fossil fuel economy, but instead a necessary, though generally unacknowledged, foundation of it. Environmental historians have convincingly situated energy use in its wider social and environmental contexts, and evaluated its impacts across society. But considerations of the role of and effects on women are also almost entirely absent from studies of environment and energy.[28]

While there have been a number of important studies of the social history of electricity, and a handful of other social histories of energy, historians generally left the study of energy to economists until the early twenty-first century.[29] Even historians of science and technology demonstrated little interest in linking the material properties of various forms of fuel or energy – a.k.a. "resources" – used by new technologies to the social, cultural, and technological changes that were their focus.[30] The majority of historians averted their gaze from the material, and turned their attention to what might be called the cultural and ideological effects of industrialization on modern societies.[31] Even though there was arguably a widespread, if tacit, acceptance that industrial societies can be recognized by a number of energy-dependent distinguishing features, these were seldom commented on explicitly, nor were they traced to changes in energy provisioning.[32] Instead, these characteristics provided the naturalized material backdrop for discussions that focused on modern human identity, culture, agency, and beliefs.

As climate change elevated awareness of the catastrophic effect that fossil fuel burning has been having on the global climate, historians renewed their interest in the Industrial Revolution. Economic historians had continued their exploration of the role of energy in industrial societies, and their work provided the foundation for a new generation of scholars.[33] The most influential of these was E.A. Wrigley.[34] In a series of books and articles published between 1962 and 2016, he revisited the idea that energy was the decisive factor in the great transformations associated with the Industrial Revolution, and rooted this in some particular material qualities of fossil fuels. His work "integrated the complex historical dynamics between demographic changes, agricultural productivity, urban growth, changing occupational structures, changes in transport facilities, and technological changes into the grand narrative on the English 'industrial revolution.'"[35]

Historians paid attention. Wrigley explained that "organic" societies (reliant on wood, water, wind, and muscle power) were obliged to rely on flows of energy from the sun, available to people in small quantities only, constrained as they were by seasonal cycles of growth, by existing high-energy demands on habitable lands (that is, competing energy demand for food, fodder, fuel, etc.), and by the difficulties and expense of transporting bulky forms of energy like horses, oxen, and wood. If an agriculturalist wanted to improve productivity by using a draught horse, for example, that draught horse would have to be fed through devoting more land to fodder, and less energy would be available for fuelwood or human food. It was a zero-sum game. Sustained economic growth was impossible. Organic economies were, therefore, typically small consumers of energy that was predominantly local. Wrigley argued that between the organic economy and the Industrial Revolution lies "the advanced organic economy": a stage Wrigley sees as promoting the division of labour and widespread commercialization of the economy. This only moderately raised the level of per capita income, but it did set the stage for the economic "take off" once fossil fuel use expanded from heating – homes, food (including breweries), and metals – to transportation and manufacturing.

The "mineral" economy of coal and later oil and gas, Wrigley argued, removed these organic limits to growth by providing dense and potent sources of power located underground, and not, therefore, competing with other land-based energies like fuelwood and food. And, most significantly, it was available to humanity as vast *stocks* of energy that people could access on demand, rather than something they depended on as a highly variable *flow* (or trickle). Once the coal-based technologies of steel and steam were realized, and efficiencies of scale were brought about through a series of positive feedback loops, these stocks of energy could be commodified, cheaply bought and sold, and moved about and used wherever demand for energy existed. The constraints previously imposed by seasonal and limited organic economies were released; fossil fuels could provide energy anywhere, and in previously unimaginable quantities, leading to sustained economic growth for the first time in human history.[36]

As one historian described it, Wrigley's account of industrialization is so widely accepted that it "now deserves the epithet of a paradigm."[37] Energy historians have welcomed its clear articulation of the role of energy as a discrete force influencing the dramatic transformations of the economy, and in ways that are clearly defined. And it nicely links the environmental climate crisis of today – an increasingly important if unintended result of the Industrial Revolution – with its cause: the burning of fossil fuels. An additional, if unremarked, strength of his approach is its integration of the household into discussions of the transition to urban industrialization. For Wrigley's work on early industrialization, which later became more explicitly focused on energy, was part of an earlier wave of research that identified the household in northwestern Europe as being a vital component of the transition from peasant economies to modern industrial capitalism, a.k.a. the transition from the organic to the mineral regime.[38]

Finding – and Losing – Women and Households in Industrialization

There is not space here to do justice to the rich historiography detailing the changing relations among women, work, and family during and after the Industrial Revolution of the 1760 to 1900 period. But in a field marked by strong and persistent disagreement, there is a general consensus that the household and women's labour – their relatively well-paid waged work in the new mills and factories, their badly paid work in most other sectors, and their unpaid work in the home – marked a significant shift in women's relationship to the economy in these years, and in ways that reshaped the emerging social, economic, and political landscape of industrial society.[39] Industrialization did not, for many households, mark a sudden break in the primary role of the household as a social and economic foundation, nor did it necessarily involve a move to a mechanized factory. Households had long produced commodities to "maximize the gross product, not the net profit" of the household.[40] Throughout the eighteenth and nineteenth centuries they continued age-old patterns of socio-economic relations, but in the changing, varied, and intensifying contexts of emerging capitalism. As Joan Scott and Louise Tilly summarized, research not only demonstrated that industrialization was "an extremely varied process," but "delineate[d] the complexity with which the family was connected to the process."[41] Research has also highlighted the integral, if varied, role of gender in the long transition from peasant to modern industrial economies. Although the exact ways that capitalism intersected with existing Euro-American household economies over time and in different places varied, there were some generalizable features over the long term: eventually most productive work was removed from the home, and waged work, much of it taken up in rapidly expanding urban areas, was eventually identified with the primary economic support of the household by a single male breadwinner. Gender played a vital role in the long transition from peasant to modern industrial economies, and new gender roles for men and women were an important consequence.[42]

But if women and the household have been identified as important components of the industrializing process, and if, as Wrigley (quoting Sieferle) and others have acknowledged, "the history of energy is the secret history of industrialisation," why are women and the household almost invisible in Wrigley's and most other energy histories?[43] One of the very puzzling aspects of women's invisibility in the history of energy is precisely the fact that women *have* long been identified by demographic, social, and labour historians (including Wrigley) as playing an important role in the transition from rural peasant societies to mature urban industrial capitalism. Furthermore, it is widely acknowledged that households stimulated the first demand for both coal heating and kerosene (paraffin) lighting.[44] Nevertheless, gender has been accorded little theoretical significance in energy histories. Histories of women in industrialization have existed in parallel to most energy-related studies of the Industrial Revolution: the role of women and households in industrialization remain on one side, energy history on the other.

Part of the problem can certainly be traced to the shadowy existence of the broader socio-economic contexts, including of labour, within Wrigley's and other economically focused narratives of energy transitions. As one critic put it, "a focus on production, exchange and the formal economy not only privileges the powerful narratives of techno-institutional supremacy as the main cause of economic growth," but in the process it "systematically silences environmental and social costs and the global inequalities incorporated into current energy regimes."[45]

> Despite forming a substantial portion of the history of industrial societies, neither atmospheric pollution, local and global, nor living and working conditions in the coalfields and factories, nor ill-health and environmental working degradation related to the extraction of mineral resources preoccupied the author [Wrigley], who completely omitted such aspects from his account of the English Industrial Revolution.[46]

While Stefania Barca, quoted here, does not explicitly discuss the role of gender or women in industrialization, the socio-ecological approach she recommends is one that can incorporate women workers, the colonial relations of which they are a part, and women's particular entanglements with the environmental consequences of fossil fuel extraction and use.[47]

Andreas Malm takes a different critical approach. He disputes Wrigley's contention that there was anything inevitable about the increase in fossil fuel use. In *Fossil Capital*, he argues that coal was adopted for manufacturing not simply because it was more abundant and easily managed than water power, but because it had some very particular advantages for nascent capitalism. The transition from water to steam power occurred in the 1820s, he argues, because the latter allowed industrialists more control over the production process, and at a time when they were in a struggle to control the men, women, and children whose labour was needed to tend and manage the new machines. As Malm emphasizes, "Anthropogenic climate change – this is part of its very definition – has its roots *outside* the realm of temperature and precipitation, turtles and polar bears, inside a sphere of human praxis that could be summed up in one word as *labour*."[48] He supplies abundant evidence of women's participation in the workforce in the early factories, a consequence, he explains, of the low wages that their work commanded. But Malm consistently describes the role women played simply as waged workers *manqué*; he takes for granted that they were unable to participate equally with men in the workforce due to some unfortunate prejudices and habits that he does not need to explain.[49] He does not identify, far less critically engage with, the *sharpened* gender roles of industrial society, where women were identified increasingly as housewives and caregivers as men were taking on their new role as full-time, family-supporting waged workers. Why and how these roles changed are questions that remain unasked in a volume that considers the changes in labour relations brought about by fossil capital to be a male preserve. This approach avoids entirely the questions, what *were* women doing in industrializing societies, and did it have anything to do with energy transitions? The answers to these questions turn out to be "lots" and "yes."

Re-gendering Industrial Society

As the extensive literature on the history of women and the Industrial Revolution confirms, women and children participated actively in the new waged forms of work, in their homes, in other people's homes, and in workshops, sweatshops, laundries, retail shops, and factories. But over time, labour became more sharply gendered in two different ways. First, the conditions of industrial waged work, and most particularly factory work, imposed new and unique burdens on mothers.[50] Many workers – men, and the employees of choice in the early mills, women and children – welcomed the relatively high wages that such labour provided, but other aspects of the first mechanized mills and factories were often greeted with intense antipathy: the noise, the heat, the noxious odours, the new forms of discipline and the unpleasant pace of factory work imposed by mechanization.[51] From the early nineteenth century, factory "hands" were coming together against employers, demanding not only increased wages but changes in a range of workplace conditions. Male workers were achieving some success by the third decade of the nineteenth century.

But the factory environment presented new and unfamiliar challenges for mothers who had previously been able to safely combine childcare with a wide range of economic activities. Working with fast-moving, automated, steam-powered factory equipment, while dangerous to everyone, was particularly difficult to combine safely with the care of infants and small children, which was part of women's culturally specific labour: as a result, "the demands of wage labour increasingly conflicted with women's domestic activities."[52] While most women did not work in

factories, over time factory work became emblematic of the problems
with married women's waged work. As Deborah Valenze details in *The
First Industrial Woman*, "Victorians were ashamed of the factory girl,
and expended much energy cataloguing her failings."[53] Factory women
were widely blamed for the remarkably high rates of infant mortality in
nineteenth-century industrializing cities. They were criticized because
they brought infants to their workplace. They were accused of neglecting
their household responsibilities and thereby jeopardizing not only the
health and safety of their families, but, by the late nineteenth century,
the future of "the race." Nineteenth-century reformers used high infant
mortality rates and household neglect to defend the increasing exclusion
of women from paid work, reformers who had little understanding of
the poverty which drove women into such difficult and dangerous work,
and which was in fact responsible for such high death rates.[54]

But it was not morality or biological imperatives that excluded
women from entry into the workforce on the same terms as men: it
was politics. The early factory system, like the coal mines, clearly did not
work for men either; many adjustments, ushered in by violence, strikes,
and the destruction of machines, had to be made before male workers
and employers could agree on the amount of hard work, danger, and risk
that male workers were willing to endure for the wages they received.[55]
These negotiations continue to the present day. The question is, why
were married women's culturally specific responsibilities for carework
and housework not accommodated at a time when the terms of men's
relation to factory work were under intense negotiation?

An important part of the answer can be found in the second
distinguishing characteristic of women's waged work: gender-
determined wages. In pre-industrial times, when wages and commodity
sales typically provided only a small portion of a household economy
rooted in subsistence, payment for women's waged labour had been
traditionally set at about half that of men.[56] Women's non-waged labour,
like men's non-waged labour, was vital to the support of the household
and the community in pre- and proto-industrializing societies, and was
recognized as such. But this traditional valuing of women's and children's
waged labour meant that they were eagerly sought out by the new industrialists, who, particularly in the first predominantly rural water-powered
cotton mills, were having a great deal of trouble attracting adult male
workers at the rates industrialists wanted to pay.[57] Men soon began to
express serious concerns about the "threat cheaper female labour posed
to the skill, status and wages" of artisans and textile workers alike.[58] They
began to advocate against women's waged work, protesting in effect
women's full inclusion in the capitalist economy as producers. Women

continued to work in factories, mills, and workshops, labouring for wages in a huge range of occupations, but they did so in jobs that were poorly paid. Their status deteriorated until it was defined as temporary, casual, and comprised of the "dead-end" kinds of jobs that still tend to be filled by young men, marginalized people, social outcasts of various sorts, and women of all ages. Historians have argued that it was toward this end – sustaining the more sharply gendered hierarchies of capitalist industrial societies – that the first protective legislation was passed limiting and sometimes prohibiting the paid work that women and children could perform.[59] Where Victorian labour reforms succeeded, in short, was in firmly entrenching women as second-class citizens in the emerging industrial capitalist system.[60] Gender was among the first of the structural inequalities that would come to define and sustain the spread of industrial capitalism around the world.

These factors – low wages, the particular and largely unaddressed challenges that waged work presented to married women workers, and male factory workers' increasing antipathy to them – emerged in the context of not only the changing nature of work, but some synergistic changes in ideas about appropriate roles for men and women more broadly. Gender distinctions were sharpening across society, shaping the new social, political, and economic relations created by fossil capital across North America and Northern Europe. Gender, in other words, was a significant factor in people's experience of the transition to fossil fuels, and it structured their roles in some new and notable ways.

Women, Energy, and Domestic Work

How have historians explained women's distinctive relationship to the transition to fossil fuels? The most significant research has, as we have seen above, occurred in the subfields of women, gender, and labour history, where it exists in parallel to energy history but seldom intersects directly with it. When feminist historians began looking at women as legitimate subjects of historical study in the 1960s and 1970s, waged labour and political involvement were among the first areas of women's experience to be examined, an emphasis that followed both from popular notions of what constituted "real" history and from the growing conviction among feminist historians that women had played an important role in the political and economic growth of their countries.[61] By the 1980s, however, the topic of women's unpaid labour in the home – housework – was receiving increasing attention from historians. Ruth Schwartz Cowan's *More Work for Mother* (1983), Susan

Strasser's *Never Done* (1982), and Caroline Davidson's similarly titled *A Woman's Work Is Never Done* (1982) were among the studies documenting the tremendous amount and value of labour that women in early industrializing countries contributed to supporting their families through their unwaged work in the home.

While seldom framed in terms of "energy history," research into women's roles and labour as housewives provides abundant evidence that women's "energy labour," whether performed as maid, mistress, or child, proved vital in converting energy from the environment into usable forms. The kinds of energy used in the home for heating, lighting, and food were originally harvested directly from the local environment, but even as money (more readily acquired by men than women) supplemented and replaced subsistence activities, it was women's labour that transformed raw materials into consumable energy – most obviously and laboriously, food energy. Perhaps more surprising, research has confirmed that housewives and their (usually female) helpers continued their energy work well into the twentieth century, powering households, and providing vital if largely unacknowledged direct and indirect energy supports to the economy and society in the process.[62]

If much of women's labour was, in an important sense, about energy production, consumption, and processing waste products, historians have demonstrated some other ways in which the household, and women's labours within it, shaped the transition to fossil fuels. As households gradually adopted commodified fuels and then foods, they were sometimes responsible for stimulating mass demand that in turn stimulated industrial growth in certain sectors. When Newcomen's first coal-powered steam engines were pumping out the coal mines to meet the ever-increasing market for coal in the eighteenth century, the initial demand was for more coal-based home heating and cooking.[63] Similarly, a substantial portion of the demand for manufactured gas and kerosene/paraffin was stimulated by the individually minute, but collectively huge, desire for more and better home lighting.[64] Increased production of these energy commodities in turn stimulated the invention of new industrial and transportation uses for this energy, which further stimulated demand, lowered energy costs, and transformed society, creating the upward spiral of economic growth distinctive of fossil fuel economies.

By the late nineteenth century, women were also widely recognized as important decision-makers regarding which kind of home energy to use for what purposes. Their domestic energy-related decisions became issues of corporate and even national interest as energy became a commodity circulating through centralized national and international networks. Women's role as energy managers not only attracted the attention of the gas, oil, and electricity companies, but a whole range of educators, reformers, and new professionals who also understood that changing energy use in the home opened up a variety of new opportunities – political, economic, financial – for men and women in modern and industrial society.[65] Notwithstanding the increasing array of consumer products and services being advertised, and the highly variable availability of gas, electric, and kerosene lighting and eventually labour-saving appliances in the twentieth century, and even as women worked for wages, it was women's muscle-power, skills, and knowledge that continued to be largely responsible for meeting the energy needs of the household, and therefore of society generally.[66]

As the work of these and other historians over the past decades has already demonstrated, gender has been an important element in the changes wrought by industrialization, and women have been active participants in energy history. The lens of energy history promises an analytical framework that could place women's home-related labours in a new light. Drawing on insights from the more contemporary "cultures of energy" research exploring the role of energy in every-day life, energy historians could situate the home at the centre of the changing energy metabolism of the nineteenth and early twentieth centuries.[67] For a focus on energy in the household not only highlights the close relationships that energy forges between people and their environments, it also urges a rethinking – or more accurately a revalu-ing – of the significance of the kind of work typically carried out by women in industrializing countries.[68]

But with some important exceptions, including of course the chap-ters in this volume, these approaches have yet to catch the attention of energy historians.[69] Explanations for why are both generational and conceptual. The term "women" remains a contested one in current historiographical debates, and the concept of "housewife" is arguably disappearing from common parlance. Over the last thirty-odd years, much-needed discussions have emphasized the diversity of women's experiences over time and geographically, and along dividing lines of ethnicity, sexuality, class, and intersectionality. Welcome attention to these factors nevertheless makes it difficult to conceptualize how powerful gender was as a category of experience, or the force with which the hegemonic designation "woman" or "man" imposed identities and behaviours on Euro-American peoples in the nineteenth and early twentieth centuries. These factors make it difficult to explain, from the vantage point of the twenty-first century, how and why the lives of a majority of women in the nineteenth and early twentieth centuries were widely acknowledged to have been "taken up largely with the domestic, that is the reproductive, the affective and the familial, and the kinds of labour that attaches most easily thereto."[70]

Much of the ongoing invisibility of the home in most written histories can be traced back to the concept of separate spheres. It continues to haunt discussions of the home. Though discredited, it continues to seep back into histories that focus on women's domestic lives, and its associations still threaten to trivialize and obscure the significance of women's lives.[71] When feminists, including feminist historians, of the 1970s and 1980s looked at women's "traditional" work in the home, what they saw were the inequalities between men, who seemed to have privileged and full access to the wealth, status, and power promised by a man's wage, and women, who largely lacked such independent wealth, status, and prestige. It is little wonder that second-wave feminists "traced [gender] inequality to the social and economic divisions between the separate spheres of family and the rest of society, particularly the world of paid work."[72]

But a much wider critique of women's inferior status had attracted significant political attention from women and men in industrializing North America and Europe from the mid-nineteenth century, and is well documented by historians. By the 1860s, feminists were already recognizing that, far from being a safe haven presided over by a nurturing mother and a firm but kind father, the home in modern industrial society was too often a locus of not only danger and disease, but "drudgery" and even "oppression." But later nineteenth- and twentieth-century reformers went further, arguing that women's inferior social, legal, and economic position could be remedied by means of a variety of strategies, from getting the vote, creating labour unions, and getting university degrees, to collectivizing housework and childcare, making homes beautiful and healthy, or later, by turning their homes into the modern, hygienic, scientifically informed spaces that men seemed to value so highly in their workplaces.[73] But by the 1980s, feminist historians were attributing gender inequality in industrial society largely to women's relegation to the domestic sphere; the solution to their inequality was equal access to paid work, or in other words, full proletarianization.

The "separate spheres" historical analysis dovetailed nicely with the contemporary problems identified by second-wave feminism. By the 1970s, the solution to women's continued economic, legal, and personal subordination was increasingly focused on the home itself: the only way for women to ensure their emancipation from mind-numbing domestic labour and from their husbands, it was argued, was for them to gain greater equality in society more generally: in order to do that, women had to renounce their domestically focused and "family obsessed" lives, and enter the workplace just like men.[74] By the 1980s, there was a broad consensus that the home, and women's unwaged work inside it, comprised the sole foundation of women's subordination. This idea has proven remarkably enduring, and remains substantially unchallenged today, forty years after historians, like North American women generally, widely embraced it.

Missing from this emancipation story, however, is the recognition of what historians such as Schwartz Cowan, Strasser, and Davidson so richly documented: the direct and vital role that women had played in providing the energy – food, light, heat – as well as the physically and emotionally demanding care work, that supported most families. There was tremendous variation in households' access to the centralized and automated networks of power becoming available across twentieth-century Europe and North America. But once acquired – as the triumphalist narratives of energy-as-liberation never tired of reiterating – it had the potential to change the ways that women worked and lived. And it did, though not always in ways anticipated by women and men at the time. Absent from the emancipation narrative is the importance of women's energy-related homework up to that point; for millennia, housework and care work was embedded in women's (and some of men's) economic work, and it really did occupy most of their waking hours. It really did support their families. Few women could ignore the demands placed on them (as servants and mistresses) to provide food, heat, lighting, water, and care for their families, or their employers' families; without them, there would be serious and immediate, as well long-term, repercussions for the household's safety, health, and well-being.

By the 1960s, and even while inequalities in energy access persisted, most European and North American women, urban and rural, were for the first time able to rely on new forms of energy to perform tasks previously accomplished only by means of their own household labour. Increasingly, they could rely on centralized networks of power to preserve and even prepare food through refrigeration and stoves, to heat their homes, and to provide cold and hot running water.[75] More women now had the time, as Schwartz Cowan has argued, to make their houses cleaner and their foods more delectable. But in those long decades of the nineteenth and twentieth centuries, after women were defined as secondary workers and identified with the domestic labour that they were tasked with performing in their homes and the homes of others, and before energy was provided through remote and automated systems, most women were tied to their homes and their families by a hundred energetic threads: timing their outings to ensure that the stove neither caught fire nor went out in their absence, checking that the lamps had sufficient oil before darkness came, performing the multiple, never-ending short- and long-term calculations involved in obtaining, preserving, and preparing food, and dealing with the waste products of food provisioning. This energy labour still remains largely invisible, and most inexplicably in energy histories.

Reducing women's domestic lives to a "separate sphere" that they need to escape from in order to become equal to men is a narrative sleight of hand that deflects attention from what women did there. Within this framework, the only real question becomes why women tolerated such a role, and, as historians have noted, framing the question *only* in this way reinforces simplistic assumptions of women's victimization and passivity. It also seems to assume that women's greater access to waged work has indeed resulted in the abolition of their domestic labour and a new equality with men. What it obscures is much of the substance of women's energy history: the economic contributions of their work, its changing social contexts and consequences, the meaning and value women and men gave to it, the alternative political economies it might represent, the expectations women had of it, their visceral experiences of energy work, and the relationships their labour created with both their environment and other people. Speaking historiographically, narratives of escape and assumptions of victimization quickly transformed into investigations of liberation; what did women do once they were "freed" from oppression in the home? Problems within the historiography, in other words, can help explain the rapid decline since the 1980s of interest in Western women's work in the home, and the limited ability of this subject to attract interest from most historians since.

Fortunately, there has long been a competing narrative around the issue of women's marginalization and increased inequality with the transition to capitalism and fossil fuels, and it has an important bearing on discussions of households, energy, and the environment. Since the beginning of the debates about women and industrialization, some feminist scholars have argued that it was not women's work in the separate sphere of the home *per se* that was responsible for women's subordinate position, but rather the *evaluation* of the kind of labour that women have traditionally done. More specifically, they have argued that the privileging of market relations in economic analysis resulted from a historically specific and politically derived (that is, not biological or essentialized) set of definitions, or evaluations, about what was kinds of activities were considered significant in society – about what "really" mattered. With the rise of industrial capitalism, women's work, paid or unpaid, almost never made it onto the list. Jane Whittle recently summarized what historians, other scholars, and activists have argued for decades now: there is simply "no logical reason for excluding unpaid housework and care work from our conception of work or the economy. Such work is necessary to the rest of the economy and does not differ substantially from other forms of work." The treatment of unpaid housework and care work "as 'non-work' has been promulgated in modern economic theory from the writings of Adam Smith onwards and have been largely unquestioned by women's and economic historians."[76]

The "separate spheres" never were descriptive categories, in other words, but rather analytical ones that functioned primarily to obscure, discipline, and marginalize women's lives. By reading women's actions and behaviours through the lens of energy history, their work in the home can provide an important counter-narrative to that of the near-invisibility of women's household-based energy, situating this work firmly at the centre of the relations that energy forms between people, between people and their environments, and between people and the economy.

Conclusion: Energy's Hidden Abode

Industrial and post-industrial capitalism have failed to deliver the emancipation that was promised women through proletarianization and wage-earning. This widely recognized failure is one of the cracks undermining twenty-first-century capitalism's claims to legitimacy, social justice, and equality, and indeed its social and ecological sustainability. Notwithstanding the hopes expressed through second-wave feminism, most women around the world continue to carry out the same domestically focused maintenance and family care work that they have always done, for pay and "for free." Other kinds of inequality are also growing, both between and within nations, and racialized and class-based lines are hardening, along with geographical boundaries.

Reflecting on the various economic, environmental, political, and social crises intensifying around the world in recent decades, Nancy Fraser notes that "we are living through a capitalist crisis of great severity without a critical theory that could adequately clarify it."[77] She locates the problem in the fact that capitalism's foregrounded claims – the benefits of market exchange, the division of labour, people's "freedom" to work for wages, and the role of economic growth in creating wealth and happiness for all – ultimately depend on "non-economic" background conditions that remain unarticulated, and indeed dissociated from the formal economy, or definitions of it.[78] These include Marx's famous condition, the role of production that lurks behind exchange, whereby workers provide the surplus value that employers extract. But Fraser summarizes the other "hidden abodes" that scholars have revealed: the role of violence and dispossession that provided access to capital around the globe, and the role of nature – the environment – in providing free materials, energetic processes, and massive sinks for waste. These are the so-called "externalities" of racialized and class-based inequality and violence to both people and non-human nature.

Fraser identifies an additional "hidden abode" of industrial capitalism: women's gendered relationship to industrial society. Fraser argues that gendered divisions are key to women's ongoing secondary status within the industrial world, and at the heart of the crisis of care now plaguing its peoples. She is not suggesting that women's "traditional" work in the home is simply the source of their oppression, to be solved by their proletarianization.[79] On the contrary, she is arguing that even as women's labour has played a vital role in creating and supporting the workforce that in turn supports capitalist industrial societies, families and the domestic sphere have in addition provided vital contributions to care, nurturing, and other kinds of support that all people require for their health and their happiness. The problem is that these aspects of life have been gendered female and restricted to the gendered space of the home, identified as apolitical, non-economic, and personal, and therefore well beyond the purview of the modern capitalist economy. The solution, she suggests, is to recognize these "feminized" concerns as social concerns, to be addressed by men and women through collective as well as individual solutions.

Gender is also a "hidden abode" of energy history. Identified as such, it provides a way of seeing and understanding the nature and value of what women contributed to energy transitions through their varied roles, even when hidden in the household. Learning how gender structured the experiences and social relations of the last energy transition puts the lives of women in the past in a new light, and might even help us understand how a new politics of gender might play a vital role in negotiating the next one.

Notes

1 Histories of women and energy in different parts of the world, and in different time periods, are also urgently needed. For a global overview of literature on women and energy from the perspective of environmental history, see Unger, "Women and Gender." Bina Agarwal's research in rural India has been particularly significant in this area. Agarwal, "Re-sounding

the Alert: Gender, Resources, and Community Action"; and "Does Women's Proportional Strength Affect Their Participation?" For a discussion of the ways in which research in contemporary development economics and feminist theories in time-use surveys can provide insights into women's labour in pre-industrializing and industrializing Western society, see Whittle, "A Critique of Approaches to 'Domestic Work,'" esp. 54–69. See as well, Harrison Moore and Sandwell, eds., *Off-Grid Empire*. Much needed is an energy history of the particular contributions of racialized women (who are not discussed in any detail here) to the energy systems of which they were a part, including those living in slave plantations, and those who contributed their waged or unwaged energy and their expertise elsewhere in the globe.

2 "The history of energy is the secret history of industrialisation." Sieferle, *The Subterranean Forest*, 137. See also McNeill, "Cheap Energy and Ecological Teleconnections." For a discussion of the lack of interest paid to the Industrial Revolution by historians from the 1990s, and the implications of this for understanding the role played by Western women in the eighteenth and nineteenth centuries, see Berg "What Difference Did Women's Work Make to the Industrial Revolution?," 22–3.

3 As we will see in detail below, there is a rich literature exploring the relationship of women to industrialization. Key works include Tilly and Scott, *Women, Work and Family*; Rose, *Limited Livelihoods*; A. Clark, *The Struggle for the Breeches*; Ross, *Love and Toil*; Lewis, ed., *Labour and Love*; Davidoff and Hall, *Family Fortunes*; and Hudson and Lee, eds., *Women's Work and the Family Economy in Historical Perspective*.

4 Both quotes are from Griffin, *Bread Winner*, 6 and 7 respectively.

5 Davidoff and Hall, *Family Fortunes*, 29. On the failure of Marxism to account for the role of the family in the class struggle, and in the culture of the working class, Jane Humphries's 1977 article is still one of the best. Humphries, "Class Struggle and the Persistence of the Working Class Family."

6 Nicholson, *Gender and History*, 11.

7 As Joan Scott famously put it, within this framework "there is nothing except the inherent inequality of the sexual relation itself to explain why the system of power operates as it does," a perspective unsatisfactory to those trying to understand the reasons behind the gendered inequality contained in these structures. J. Scott, "Gender: A Useful Category of Historical Analysis," 1,058.

8 Whittle, "A Critique of Approaches to 'Domestic Work'" 36.

9 Griffin, *Bread Winner*, 5. As Griffin goes on to add, where women "have been less successful though has been inserting the domestic into the mainstream ... which remain[s] focused upon their traditional concerns of politics, empire and prosperity," 5.

10 Whittle, "A Critique of Approaches to 'Domestic Work,'" 70. The unequal distribution of wealth within the household, as well as particularly low

wages for women, had attracted the attention of social reformers such
as Charles Booth and B.S. Rowntree from the late nineteenth century.
By the early twentieth, women were actively involved in documenting
this inequality. The historians mentioned above were among those
who made excellent use of the research conducted by sociologists such
as Maud Pember Reeves, *Round About a Pound a Week*, and Eleanor
Rathbone, *The Disinherited Family*.

11 Griffin, *Bread Winner*, 61. It would be difficult to find a historian
of women's domestic lives and work who has not explicitly made
this point.

12 These definitions are from Greenberg, "Energy Flow in a Changing
Economy," 31. There has been considerable discussion, and little
agreement, about the meaning of "energy." See Lindsay, *Energy:
Historical Development of a Concept*, and Zachmann, "Introduction." For
a very different look at the history of energy as discourse, see Daggett,
The Birth of Energy.

13 Crosby, *Children of the Sun*, esp. 6–24.

14 A "horse power" was a particular kind of machine. It was basically a
treadmill, with sides attached, into which a horse was put to provide
power. By walking on the treadmill, the horse turned a rod, which was
attached to a gear, which was in turn attached to a machine. Smaller
"dog powers" were also used.

15 Greenberg, "Reassessing Power Patterns in the Industrial Revolution,"
1245; Smil, "Energy Flows in the Developing World"; Colpitts,
"Food Energy and the Expansion of the Canadian Fur Trade"; Wynn,
Canada and Arctic North America, 113. The expression "endosomatic" is
McNeill's, described in *Something New under the Sun*, 11.

16 Miller and Warde, "Energy Transitions as Environmental Events," 466.
The authors are describing Warde's "The Hornmoldt Metabolism." Karl
Marx alluded to this phenomenon in *Capital*, which was the basis for
later work on the metabolic rift. Marx, *Capital*, vol. 3, 949.

17 The implications of the invisibility of modern energy is emphasized by
a number of historians, including French, *When They Hid the Fire*, and
Johnson, *Carbon Nation*.

18 Historians such as Manuel González de Molina describe human
epochs as "metabolic regimes," characterized by the increasing
appropriation of material and energy flows, and David Christian
demonstrates the consistency of human energy intensification in
the very long run, noting that humans, like all organisms, are "anti-
entropy machines." De Molina and Toledo, *The Social Metabolism*, 155,
267–75; Christian, *Maps of Time*, 80. See Sandwell, "An Introduction to
Canada's Energy History," for a different approach.

19 Historians are now creating much more complex discourses about energy transitions, but for a recent manifestation of this triumphalist heroic and entirely masculine narrative, see BBC's three-part television series *Shock and Awe: The Story of Electricity*.

20 Greenberg, "Energy, Power, and Perceptions of Social Change," 694; Zachmann, "Introduction," 8–22; Barca, "Energy, Property and the Industrial Revolution Narrative." One of the most famous early dystopian visions of what the new kinds of energy would bring was Mary Shelley's *Frankenstein, or the Modern Prometheus* (published in 1819). For discussions of nineteenth-century energy anxiety and the demonization of coal and steam, and for the "dark side" of electricity narratives in particular, see Gooday, *Domesticating Electricity*, and Simon, *Dark Light*.

21 Toynbee, *The Industrial Revolution*; Ostwald, "The Modern Theory of Energetics," 510–11, cited in Reynolds, *Stronger than a Hundred Men*, 2. For the positive holistic influence of Ostwald's energetics on environmental awareness, see Tanner, "Thinking with Energy," though she does not mention that women's activities were "counted" in his more holistic view of the role of energy throughout society. Perhaps most remarkable among the utopian visions of new kinds of energy to transform society is John Adolphus Etzler's *The Paradise within the Reach of All Men, without Labour, by Powers of Nature and Machinery*, from 1833. He envisioned a world abundantly powered by mostly renewable resources, including arrays of hundreds of mirrors reflecting light from the sun. See also Greenberg, "Energy, Power and Perceptions of Social Change," 694–6.

22 Cottrell, *Energy and Society*, 2. Leslie White was also influential in reiterating these theories, for example in "Energy and the Evolution of Culture" and *The Science of Culture*, esp. 362–93.

23 Cipolla, *The Economic History of World Population*, 35–6.

24 Merchant, *The Death of Nature*, and *Earthcare*.

25 Bernstein's "On Mother Earth and Earth Mothers" provides a particularly succinct summary of these contemporary debates. See as well, Adams and Gruen, *Ecofeminism*.

26 Debeir, Deléage, and Hemery, *In the Servitude of Power*; Chapman, *Fuel's Paradise*.

27 Notable as early energy histories are Melosi, *Coping with Abundance*; McNeill, *Something New under the Sun*; and Cronon, *Nature's Metropolis*. Williams's *Energy and the Making of Modern California* is perhaps the first book-length environmental and social history of energy, a case study of each kind of energy used in Californian society.

28 On the absence of women in environmental and energy history, see Leach and Green, "Gender and Environmental History"; Haraway, "Anthropocene, Capitalocene, Plantationocene, Chthulucene"; and Unger, "Women and

Gender." There have been some important studies of women, health, and
the environment. See, for example, D. Scott, ed., *Our Chemical Selves*; and
Murphy, *Sick Building Syndrome*.

29 These important social histories of electricity include Hughes, *Networks
of Power*; Nye, *Electrifying America*; and Tobey, *Technology as Freedom*. The
twenty-first century has given us many more of these environmental social
histories, including Mosley, *The Chimney of the World*; Zallen, *American
Lucifers*; Sandwell, "The Coal-Oil Lamp"; Johnson, *Carbon Nation*; Adams,
Home Fires; Hecht, *Being Nuclear*; and K. Brown, *Plutopia*.

30 For an exception to this generalization, see Zimring, *Aluminum Upcycled*.

31 For recent critiques of cultural constructionist interpretations of "nature,"
see Malm, *The Progress of This Storm*, 21–43; and LeCain, *The Matter of
History*. Daggett, *The Birth of Energy*, provides a good example of the
constructionist approach.

32 With thanks to Steve Penfold (personal correspondence) for articulating
these features so clearly: near or total energy invisibility at the point of
consumption; production/extraction in mega-systems with brute-force
technologies; a centralized and hierarchical administration (whether
publicly or privately controlled); cheap, fast, and extensive transportation
and communication; the wide dispersion of new energies into multiple areas
of social, economic, and political life; and almost exclusive reliance on non-
human (nonsomatic) energy.

33 Some energy historians were much more critical of the dominant
triumphalist narratives of progress. Roger Fouquet, in his detailed studies on
the relationship between economic development and energy use over the last
centuries, explicitly included environmental damage and social inequality
in his evaluations of energy transitions. See, for example, Fouquet and
Pearson, "Seven Centuries of Energy Services"; Fouquet, "The Slow Search
for Solutions"; and Fouquet and Broadberry, "Seven Centuries of European
Economic Growth and Decline."

34 Wrigley, "The Supply of Raw Materials in the Industrial Revolution";
Wrigley, *Energy and the English Industrial Revolution*; Wrigley, *The Path
to Sustained Growth*.

35 Saelens, "Review of *The Path to Sustained Growth*."

36 For a critical analysis of the ideology behind constructing capitalist
industrialization as a "release" of "natural constraints," see Wood, *The Origin
of Capitalism*.

37 Malm, *Fossil Capital*, 24.

38 The work of the Cambridge Group was particularly important in adopting
the meticulous demographic work of John Hajnal on the northwestern
European marriage pattern to explanations of industrialization. See, for
example, Laslett, *The World We Have Lost*; Wrigley, *Continuity, Chance and
Change*; and Levine, *Reproducing Families*.

39 Some classic works include Pinchbeck, *Women Workers and the Industrial Revolution*; Hewitt, *Wives and Mothers in Victorian Industry*; Berg, *The Age of Manufactures*; Tilly and Scott, *Women, Work and Family*; Reddy, *The Rise of Market Culture*; Valenze, *The First Industrial Woman*; and Hudson and Lee, eds., *Women's Work*. Detailed case studies documenting the value of women's labour to the process of proto-industrialization and full industrialization include Gullickson, *The Spinners and Weavers of Auffay*; Lehning, *The Peasants of Marlhes*; Accampo, *Industrialization, Family Life and Class Relations*; and Gray, "Rural Industry and Uneven Development." For critical summaries of this literature, see Clark, *The Struggle for the Breeches*; Hartman, *The Household and the Making of History*; Whittle, "A Critique of Approaches to 'Domestic Work'"; and Berg, "What Difference Did Women's Work Make to the Industrial Revolution?," 26.

40 For a discussion of the limited role of steam power and the rapid increase in both population and traditional labour power in creating the "industrious revolution" prior to the 1820s, see de Vries, *The Industrious Revolution*; Greenberg "Reassessing Power Patterns of the Industrial Revolution," esp. 1241; and Samuel, "Workshop of the World." For a discussion of the wider European context see the classic work by Kriedte, Medick, and Schlumbohm, *Industrialization before Industrialization*.

41 Tilly and Scott, *Women, Work and Family*, 1.

42 Gray, "Rural Industry and Uneven Development." For a synthetic overview of gender, women's labour, and domestic work, see Whittle, "A Critique of Approaches to 'Domestic Work'"; and Horrell and Humphries, "The Origins and Expansion of the Male Breadwinner Family."

43 Sieferle, *The Subterranean Forest*, 137. Also quoted by Wrigley, *The Path to Sustained Growth*, 31.

44 Jones, "The Carbon-Consuming Home." On the importance of coal to household practices, see R. Allen, *The British Industrial Revolution in Global Perspective*; H. Harris, "Conquering Winter"; S. Adams, *Home Fires*; Mosley, *The Chimney of the World*; Cavert, *The Smoke of London*; Malm, *Fossil Capital*; and Brewer, *From Fireplace to Cookstove*.

45 Barca, "Energy, Property and the Industrial Revolution Narrative," 1309.

46 Ibid., 1310. On the limitations of classic economic theory identifying energy itself as something worthy of economists' attention, see Kander, Malanima, and Warde, *Power to the People*, 6–7.

47 There is a growing body of scholarship documenting the extent to which the Industrial Revolution resulted in diminished health for many in the industrializing Western world. See Floud, et al., *The Changing Body*; and White, *The Republic for Which It Stands*, chapter 3, "Dying for Progress."

48 Malm, *Fossil Capital*, 6.

49 Malm notes the ongoing importance of subsistence support, which continued to be important to households, along with wages, well into

the nineteenth century. The importance of women's work in this regard is implied rather than stated. Malm, *Fossil Capital*, 137–8. Women are mentioned twice in the index: "work in mines" and "work in factories" (both references to page 238).

50 Tilly and Scott, *Women, Work and Family*; A. Clark, *The Struggle for the Breeches*; Valenze, *The First Industrial Woman*; and Berg, "What Difference Did Women's Work Make to the Industrial Revolution?" Each has a slightly different explanation of why these distinctions emerged, but agree with the other scholars in the field that sharpened gender distinctions were a defining characteristic of early industrialization. For descriptions of women's reflections on their factory labour, see Griffin, *Bread Winner*, 42–52.

51 There is an extensive literature on workers' responses to the early factories. See, for example, Thompson, *The Making of the English Working Class*; Reddy, *The Rise of Market Culture*; Mosley, *The Chimney of the World*; Malm, *Fossil Capital*; Montrie, *Making a Living*; Griffin, *Bread Winner*.

52 Tilly and Scott, *Women, Work and Family*; Hewitt, *Wives and Mothers in Victorian Industry*; Griffin, *Bread Winner*.

53 Valenze, *The First Industrial Woman*, 3. There is a large literature examining the extent and the causes of the high infant mortality rates in nineteenth-century industrializing cities in Europe and North America, well summarized in A. Clark, *The Struggle for the Breeches*. For a discussion of the growing conflict between women's domestic and paid work, and the late nineteenth century association of mothers' work with racist ideas about the "future of the Empire," see Davin, "Imperialism and Motherhood."

54 Hewitt notes how the efforts of reformers were thwarted by their emphasis on women's morality, and not on the problems of working women, and the poverty which created the need for their work. Hewitt, *Wives and Mothers in Victorian Industry*. The work of sociologists and historians documenting over the past hundred and more years the appalling living conditions, and high rates of infant and child mortality, of so many working-class families in industrializing nineteenth-century cities is gaining substantial support from recent quantitative medical research. See, for example, Floud, et al., *The Changing Body*.

55 Thompson, *The Making of the English Working Class*; Malm, *Fossil Capital*; Andrews, *Killing for Coal*; Mitchell, *Carbon Democracy*; Johnson, *Carbon Nation*.

56 Pinchbeck, *Women Workers and the Industrial Revolution*, 2. On the ongoing impact of very low wages on working-class women's work, their independence, and the way that poverty shaped their lives, see Griffin, *Bread Winner*, esp. 51–61. On the relationship between low wages and the extent of women's factory work, see Berg, "What Difference Did Women's Work Make to the Industrial Revolution?"

57   Valenze, *The First Industrial Woman*; Boydston, *Home and Work*; Malm,
     Fossil Capital.
58   Clark, *The Struggle for the Breeches*, esp. chapter 7, "The Struggle over the
     Gender Division of Labour, 1760–1826," 139; Scott, J., *Gender and the Politics
     of History*; Griffin, *Bread Winner*.
59   See, for example, Gray, "Factory Legislation and the Gendering of Jobs in
     the North of England"; Rose, *Limited Livelihoods*; and Backhouse, *Petticoats
     and Prejudice*.
60   The "dual economy" theory argues that there are structural aspects of
     capitalism that require the type of labour that women perform in the
     workplace. Stewart, *Women, Work and the French State*. On the decreasing
     status of women's labour in the eighteenth and nineteenth centuries, see,
     for example, Boydston, *Home and Work*; and Schwartz Cowan, *More Work for
     Mother*. One of the most recent powerful articulations of women's second-
     class status inside the working-class family as well as outside can be found
     in Emma Griffin's *Bread Winner*. Supplementing early-twentieth-century
     household studies (for example, Reeves, *Round about a Pound a Week*, first
     published in 1913, and Rathbone, *The Disinherited Family*, first published
     in 1924), Griffin's examination of nineteenth-century working-class
     autobiographies explores not only women's limited ability to earn a living
     wage, but the limited extent to which men's overall rising standard of living
     in nineteenth-century working-class Britain was shared with their wives and
     children, and women generally. Griffin, *Bread Winner*.
61   On the long battle of the 1960s and 1970s to include women as both
     legitimate historical subjects and professional historians, see Lerner, *Living
     with History/Making Social Change*.
62   See, for example, Baillargeon, *Making Do*; Luxton, *More than a Labour of Love*;
     Hagemann and Roll-Hansen, *Twentieth-Century Housewives*; Strong-Boag,
     The New Day Recalled; Clear, *Women of the House*; Ross, *Love and Toil*; Lewis,
     ed., *Labour and Love*; and Parr, *Domestic Goods*. Historians have also been
     re-evaluating the importance of women's cleaning and hygiene labour in
     gradually improving the health of late-nineteenth- and twentieth-century
     households, in the face of contagious diseases, and increasing coal- and
     oil-based environmental pollutants. Women's particular responsibilities
     as energy and care providers made them active participants in shaping
     domestic architecture, and in influencing the design of rapidly growing
     cities. A. Adams, *Architecture in the Family Way*; Kiechle, *The Smell Detectives*;
     Tomes, *The Gospel of Germs*. Hayden, *The Grand Domestic Revolution*. We
     can anticipate more studies of health and hygiene, and women's role in
     promoting them, in the aftermath of the covid-19 pandemic.
63   The same was true in the United States, where demand for home heating
     provided the initial impetus behind anthracite coal extraction and

transportation, lowering the cost of coal until it became an attractive alternative to wood for industries. Jones, "The Carbon-Consuming Home"; Sandwell, "People, Place and Power"; Mosley, *The Chimney of the World*, 18–19; R. Allen, "The Shift to Coal," 12; Cavert, *The Smoke of London*; Malm, *Fossil Capital*. Wrigley notes that in London by 1700, coal had largely replaced wood for home heating and cooking: "the switch from wood to coal … enabled roughly 800,000 acres of woodland to be used instead to produce food, or wool and hides," stimulating waged labour, manufacturing and transportation industries, and urban consumerism, all leading to economic growth. Wrigley, *The Path to Sustained Growth*, 17.

64   Sandwell, "The Emergence of Modern Lighting in Canada"; O'Dea, *The Social History of Lighting*; Eckirch, *At Day's Close*.

65   Goldstein, *Creating Consumers*; Tobey, *Technology as Freedom*; Nye, *Electrifying America*; Clendinning, *Demons of Domesticity*; Sandwell, "Pedagogies of the Unimpressed."

66   Even as more prepared foods became available for purchase in major cities from the mid-nineteenth century, research suggests that most food continued to be prepared and consumed "at home" until the mid-twentieth century. In the Canadian context, see for example, Iacovetta, et al., *Edible Histories, Cultural Politics*; and in the British, Griffin, *Bread Winner*.

67   Shove and Walker, "What Is Energy For?" See, for example, research and publications associated with the Material Cultures of Energy project, http://www.bbk.ac.uk/mce; and Trentmann, ed., *The Making of the Consumer*.

68   This deep inclusion of women's work into environmental history can be seen clearly in Nash, *Inescapable Ecologies*; Valencius, *The Health of the Country*; and Montrie, *Making a Living*. On the importance of health in discussions of industrialization, see White, R., "Dying for Progress"; and Floud, et al., *The Changing Body*.

69   Important exceptions include Gooday, *Domesticating Electricity*; Jones, "The Carbon-Consuming Home"; and Harrison Moore and Sandwell, eds., *Off-Grid Empire*.

70   Lewis, ed., *Labour and Love*, "Introduction." In her famous 1986 essay, "Gender: A Useful Category of Historical Analysis," Joan Scott articulated the need to understand gender as a category of analysis, as a way of understanding dynamics of power (pun intended) in particular societies in particular times. I am recommending this approach here. There has been a great deal of research as well into the power of hegemonic masculinity since the industrial era. See, for example, Sarti, ed., "Men at Home"; and Huebel, *Fighter, Worker, and Family Man*.

71   See McGraw, "No Passive Victims, No Separate Spheres."

72   Sandwell, "The Limits of Liberalism," 429.

73   For examples of these broad women's reform initiatives for architecture,

home economics, hygiene, and labour respectively, see Hayden, *The Grand Domestic Revolution*; Sandwell, "Pedagogies of the Unimpressed"; Tomes, *The Gospel of Germs*; and Hewitt, *Wives and Mothers in Victorian Industry*. There is also a significant literature about women's political reform movements of the nineteenth century in the countries of North America and Europe, as well as about the late-nineteenth-century increase in women's professionalization, including in engineering, law, and medicine.

74 This is not to deny the inequality and oppression, identified since the nineteenth century, inherent in the bread-winner model. For a powerful critique of its mid-twentieth-century manifestation, see Friedan, *The Feminine Mystique*. The twenty-first century (and exacerbated by the COVID-19 pandemic) witnessed a new romanticizing narrative: a new celebration of domesticity and women's "traditional" labours. As critic Jennifer Bernstein summarized, "While the new domestics advocating home brewing, fermenting kombucha, and churning butter are likely aware of their irony in an era of unprecedented technological progress, this nostalgia does little to further the goals of middle- and lower-class women in the developed world." Bernstein, "On Mother Earth and Earth Mothers."

75 Most historians have accepted Schwartz Cowan's suggestion in *More Work for Mother* that modern appliances simply increased the domestic workload for women by increasing standards. As this volume argues, it is time to reassess the extent and value of women's home-based energy work. In an econometric study of US census data, Emanuela Cardia found that women's labour force participation correlated with the existence of modern appliances in their mid-twentieth-century homes, but most particularly by having indoor running water. Cardia, "Household Technology." Bernstein, and Bowen and Elliott remind us that these tasks have not entirely disappeared with the advent of modern energy, and that they continue to weigh most heavily on the poor. Bernstein, "On Mother Earth and Earth Mothers"; Bowen and Elliott, "Joy of Cooking?"

76 Whittle, "A Critique of Approaches to 'Domestic Work,'" 54.

77 Fraser, "Behind Marx's Hidden Abode," 56.

78 Ibid., 57–8.

79 In this she differs from Illich, who sees no meaningful experience in the role of "the housewife" in modern capitalist societies. Illich, *Shadow Work*.

2

Finding Women in the History of Lighting: Candles in the English Home, 1815–1910

Karen Sayer

For either in the Morning, or at Night,
We piece the Summer's Day with Candle-light
Mary Collier, "The Woman's Labour: An Epistle to Mr Stephen Duck, in answer to his late poem, called The Thresher's Labour" (London, 1739)

In winter, get the work forward by daylight, to prevent running about at night with candles. Thus you escape grease spots, and risks of fire.
Enquire Within Upon Everything (London, 1894)

Introduction: Looking for Women in the Histories of Candlelight

There is a well-worn technocratic tale of the steps humanity has taken after dark: moving from dependence on moonlight and firelight, to illumination by rush-candle, tallow, wax, and lamplight, then by gaslight, and finally by arc and incandescent electric light. There are established vignettes within that tale designed to demonstrate these advances through examples of light generated for the purposes of public display: for celebration, show, and amusement – for coronations, shop windows, or in pleasure gardens – its harnessing to the municipal causes of governance and safety, navigation, capital, and production. The positivist story of artificial light – as told by enthusiasts for each new technology long before they were written into history – so often framed by public life and civility, in and of itself represented the essence of Western progress through its dominant colonial association with "civilization."[1] In focusing on candles as a continuous yet complex technology of illumination to reframe histories of illumination as entangled, complex, and interconnected, we should also be reminded to consider the interconnections and energy flows involved in making and using light: the whales and whalers, the miners, the sheep and cattle, the shepherds and graziers, the men and women who got up before dawn and retired after dusk while employed in husbandry, managing young animals, milking, herding, and harnessing, and the women who were paid to make gas mantles and electric light bulbs, who were cooks, chambermaids, nursery nurses, and housemaids, and those who cared night and day, unpaid, for children, the sick, disabled, and elderly. The history of energy consumption should never be read as linear, but as grounded in the ordinary complexity of everyday life. Looking for women in the history of lighting, we find that the story shifts away from a narrative of progress and towards a much more entangled picture of intricate social relations and decisions about energy use, embedded within the paraphernalia of nineteenth-century life as it was lived.

Looking at the social history alone, we see that the social identities implicated in the production, reception, and use of energy were actively created – they were not simply imposed. The use of candles, oil, and gas, for instance, involved the production of goods to ignite the fuel, such as Lucifer matches, still relatively new and fascinating enough in the mid-nineteenth century to merit poetic reference,[2] while the "match girls" who made them famously went on strike in 1888 at Bryant and May. We ought also to address the production of the buildings, equipment, and goods needed to generate and hold the sources of light. In the case of candles, and their paraphernalia, we see object collections full of rush-light holders, lanterns, and candlesticks, from those that were pocket-sized for travel to practical designs that caught the wax as you walked to bed, or decorative mantelpiece objects, through to the most decorative, multi-tiered chandeliers, made variously of wood, glass, china, papier-mâché, copper, brass, iron, pewter, and more:[3] cultural and material expressions of energy and its use. At the level of fuel, there was innovation (and efficiency gains to be purchased for those who could afford it) in the form of new wicks, different types of wax, fat, oil, and additives. Beeswax was still used, but in the first half of our period, wax and oil predominantly came from whales ("spermaceti"), land animals, especially sheep and cattle, and vegetable sources, such as linseed, rapeseed, coconut, palm kernel, groundnuts, cotton seeds, soya, and olive, to which was added the mineral/fossil fuel–derived paraffin oil.[4] In the nineteenth century, candles were made variously of beeswax, tallow, paraffin wax, stearic acid and sperm whale oil, palm or coconut oil, or combinations of these materials, all of which had to be rendered and processed. Some burned more efficiently than others – tallow, the cheapest fuel, burned least cleanly. The cheapest "candle" of all was the rush-light, made from peeled rushes dipped in fat, with the pitch acting as a wick, but most tallow candles were made by local chandlers. As William O'Dea noted in his classic account, the *Social History of Lighting* (1958), based on the object collection he curated from 1935 at the Science Museum in London, anything oily or fatty could be used for lighting, and even birds were used as sources of fuel.[5] However, the qualities admired in the most expensive candles were their "brilliance" (white light) and that they required no snuffing (that is, they required no wick trimming while in use). These were made by the larger manufacturers who, by investing in new equipment, sought to make the most of materials such as palm oil and paraffin, in order to produce so-called patent candles, such as "Palmitine" candles – a field in which there was intense competition.[6]

The production of these so-called patent candles was expensive, requiring specialist equipment to refine the raw materials, as well as dangerous. In 1846 a fire at Messrs. Palmer, "the patent candle manufacturers" in Derby, for example, was estimated to have caused over £50,000 worth of damage through losses in stored candles, raw materials, and "steam machinery," despite the swift action of "parochial fire engines" and police.[7] Given the exigencies of demand and profits to be made, arsenic was used in the manufacture of some stearic acid and spermaceti candles, in order to improve the colour of the wax and to create light of extra brilliance and purity, which carried its own commercial risks. According to a report in the *Times* (1837), the proportion was "1 part white arsenic to 27 parts fatty matter." Reassuringly for readers, the presence of arsenic could apparently be detected as a smell of garlic when the candle was blown out. When this was initially discussed no one was quite sure if it was necessarily (medically) bad – some reported no problems using these candles, while "others of weaker habit of body had found them injurious." However, as they were often sold as (more expensive) wax candles, the *Times* was quite sure that a "fraud is being practiced on the public."[8] The issue resurfaced in 1873 when a number of different types of green candle, including Christmas candles, were tested to see if they contained arsenic as a colourant. Most of those tested were coloured by verdigris or ultramarine green, but some tapers did include arsenic. This type of taper was normally used to light gas, but the *Lancet*, from which the report was drawn, expressed concern about any use of arsenic in candles, especially on Christmas trees.[9] Coloured candles on a Christmas tree were pinpointed as the cause of the poisoning of a number of children at a party in 1889 – every eight green candles equaled one grain of "arsenious anhydrite," and as the red candles were coloured by vermilion (made from cinnabar, that is, mercury sulfide), a degree of mercurial poisoning was added to the mix.[10]

The risks associated with arsenic therefore applied across the period (and were especially bad in crowded environments, where many candles would be burning at the same time).[11] Reports such as these, while newsworthy and thus potentially the exception rather than the rule, nevertheless incidentally show us where and how candles were being used, such as for decoration and celebration, as well as for practical purposes. Indeed, despite the risks and attendant costs, and despite technologically innovative competitors (in new forms of oil lamps, as well as gas and electricity), candles remained such an adaptable, vital, and profitable form of illumination throughout the period that companies like Price's (a competitor of Palmer's) still considered it to be well-worth investing in a new plant and new machinery, even at the end of the century. Candles manufactured in the UK were still being sold in the domestic market at the end of the nineteenth century, while markets available to these same manufacturers, entangled with the expansion of empire, were likewise expanding. Hence Price's noted in their 1884 report to shareholders (the thirty-seventh since being established) that they "had to contend against great opposition … they could not raise the price of the candles, so they could only reduce the cost of manufacture as much as possible, and be content with small profits and a large business." However, "[t]rade had been healthy," they reported, and they were also the winners of a gold medal at Melbourne. The result was a healthy dividend,[12] and their total value of candle exports in the 1870s had been increasing with the expansion of colonial markets.[13] This shows us how energy persists: its manufacturers adapt to new techniques and markets, meanwhile the cultural meanings attaching to old forms of energy – as well as their consumption – expand into new purposes and uses.[14]

As can already be seen with the example of candlelight, the history of energy, therefore, contains more continuity and more complexity than the original teleologically framed nineteenth-century histories of illumination might suggest,[15] as more recent social and cultural histories argue.[16] Moreover, despite the suggestion that "[s]o basic and unremarkable an activity as domestic lighting has left few records and excited little contemporary comment,"[17] it will become clear in this chapter that this history can be teased out from the incidental evidence in sources such as the local and national press, as well as advisory texts, periodicals, and, occasionally, testimony. The meanings attached to illumination as paradigmatic of human progress are well known. Both Schivelbusch and Nead, for example, make reference to Bachelard's writing on the lamp. In Schivelbusch we see the following quotation: "the lamp is the spirit that watches over every room. It is the centre of the house. A house without a lamp is as unthinkable as a lamp without a house."[18] "For Bachelard," Nead writes, "the lamp is humanised and the experience of lamplight is an intense, psychological relationship. Time illuminated by the lamp is solemn and observes a slower tempo than the temporality of daylight or electric light. It is a period of dream, contemplation and reverie." It is, Nead observes, what Bachelard calls "igneous time," that is, "time that passes by burning."[19] The lamp and the individual are in a relationship; for Bachelard the lamp waits, watches, and guards.[20] However, Nead and Schivelbusch interpret Bachelard differently when exploring the transition between lamplight and gaslight (which Bachelard does not mention). For Schivelbusch, there is a sharp transition between them, in which the shift to gas-lighting, produced industrially, is likened to the shift from coach to railway – gaslight and the railway often having been likened to each other in the nineteenth century. He also suggests that people were made uneasy by the infiltration of their homes by industry, and the loss of freedom that was attached to this. People could no longer become lost in the contemplation of the "'individual' flame of an oil-lamp or candle"; no one, he believed, gazed into a gas flame.[21] For Nead, however, the move is more subtle, and she extends Bachelard's work so that "gaslight can be understood to occupy a place somewhere between the archaic form of the lamp flame and the crude modernity of the electric light."[22] In part, the difference in interpretation is down to Schivelbusch's focus on domestic light, and Nead's focus on street light. But this raises important, wider questions about the perception not just of night and day, but also of different qualities and types of light, and the framing of the histories that are told of them. The history of candlelight is framed as inherently ancient and archaic, and its continuity, as well as its "domestic" history, is perhaps simply less spectacular than the histories sought by positivist, normative narratives of linear "progress" and economic dominance gained through the successful opportunity costs of technological innovation, especially when they take place in public spaces. Linear narratives by default recycle the paradigms used within the rhetoric of civilization that dominated nineteenth-century languages of light. In this feminist historiographical sense, the histories written of illumination have been deeply gendered all along.

As feminist historians posited during the 1980s and 1990s, there was (and remains) a fundamental structural issue with the production of historical knowledge that has resulted in the exclusion of women from the record. This was due to the form of the subject as much as its content. It is not, they argued, that we must simply bring women into the field of historical inquiry, for instance by reminding ourselves that there *were* significant individual women ("women worthies," as it were) in the past, or that women as well as men participated actively in the nation's economic or political life. Rather, they argued, we need to shift history as a field of inquiry so that historians will always recognize that history includes women, and women's experiences, and that their work, though historically assigned a low status, should be recognized as being of equal importance, recognized for the skill and intelligence it required, and given its due material value and creative worth.[23]

Another approach to rectifying this situation is methodological. As part of this broader undertaking to restructure the field of history, feminist historical practice frequently necessitates addressing the gaps within and looking to the margins of the available sources to find women's histories – in this case, of candlelight. Some of these women were written about and occasionally interviewed by the authorities in official inquiries, or appear via incidental references and panegyrics to ideal motherhood in other people's autobiographies. We catch similar glimpses of them or their household tasks in descriptive and prescriptive sources such as articles in periodicals, or the national and local press, or in popular advice texts printed by companies like Sunlight Soap (which aimed to advertise as much as instruct) and housewives' manuals such as *Mrs Beeton's*. The latter went through multiple editions and emphasized that managing domestic life and making the home comfortable was a woman's first and foremost role – if often downplaying the work required.

For instance, the *Sunlight Year Book* of 1898 suggested that the "number of things that may be done to make home pretty, agreeable, and tasty, and to make the best of what you have, are almost inconceivable," thereby setting a tone typical for the period about domestic comfort and thrift, through a discussion that is clearly gendered. After providing a number of examples (including knitted socks, pincushions made from a cigar boxes, furniture from hampers, "a music wagon" from a box, and an umbrella stand from a "drain-pipe cleaned and painted"), the article suggests that although these activities require "taste and intelligence," this is (simple and easy) women's work. "All these things," it states, "may be done by the ladies of the house, for they are not arduous work." These suggestions were made on an explicitly prescriptive basis, which, it implied, it delineated for the good of the household's finances and the benefit of the housewife's mental health as manager of the following: "The Right Way – A great source of domestic economy is in learning to do things in the right way. There is a right and a wrong way of doing even the simplest things. The right way saves time, trouble, temper and money; the wrong way does the reverse."[24] We know that these types of books were used (and that their readers did not necessarily accept them at face value) because their owners occasionally annotated them. The particular edition of the *Sunlight Year Book* that was examined for this chapter has "Most useful book" written on the outside of the front cover, and notes on helpful pages written inside the front cover – such as "'Medical' Turn to page 368," and "Leach" written over "rest in bed is necessary" in the entry for influenza, as well as "sunlight soap" being crossed out and "Watson's Soap" written in, all of this done in fountain pen, through which we see a reader's hand at work, and critical reflection in play.[25]

In other words, to frame our enquiries in such a way as to give weight to women's lives, especially to those who were working class, or unable to write, and to the intersections of their experiences, we must write using the interstices and look beyond the official written record to sources that are ephemeral, incidental, accidental. This is not to say that the fundamental tools of historical practice change, but that the questions alter. As Catherine Hall put it in her lasting account of feminist historical research methodology, *White, Male and Middle Class*, the "stories which [historians] construct, from laborious archival work ordered by conceptual frameworks, are grounded through an attempted comprehensiveness in relation to evidence, a commitment to look at countervailing accounts, an effort to test interpretations against others, the practices of good scholarship [remain unaltered but] ... [t]he meaning of being an historian over the last twenty years, of trying to do certain kinds of historical work, has significantly changed."[26] The commitment to this method, but also to changing "certain kinds of historical work," has continued. From the 1960s and 1970s, feminists and feminist scholars undertook historical work across a wide range of inquires that, by the late 1980s and 1990s, most strikingly included histories of previously omitted topics related to private, domestic life and its objects: sewing, housework and cleaning, washing machines, and cookers. This is what historians such as Melanie Kiechle have been doing, in accounts that ensure that women's specific role, the societally bounded exercise of gendered knowledge and expertise within the home in the West, is made visible in any environmental analysis of past interactions of new "human" technologies and environmental conditions.[27] Looking at candlelight in the nineteenth century reminds us that, as Virginia J. Scharff has said, "humans know nature, at least in part, through gender."[28]

To understand decisions regarding energy use and candlelight, I would argue, we need to understand the history of decision-making and how the choice to use a form of energy other than that of manual labour became habitual, and that those decisions were inescapably gendered, in the way that the rest of Victorian life was at least prescriptively gendered.

Because we urgently need to rethink our decision-making in everyday life in the context of global climate change, and as Haraway argues to recognize the embeddedness of humanity – and our species' place – within the environment,[29] we need to understand the history of that everyday decision-making, and make its unseen continuities within the ordinary visible and thereby open to discussion. Looking at the energy of candlelight involves an environmental reading of *habitus* – a history that seeks to understand how people have embodied routines of thought and practice passed to them by the social networks and worlds that they inhabited. This takes us into an environmental history of dwelling,[30] the ways in which people have lived their day-to-day lives, their interconnectedness with each other and the rest of the environment, and their commonplace reliance on energy. When environmental history touches on the history of energy, and the history of energy (as it must) touches on its use in the home (an environmental history of small, everyday things, connections, and relationships), then that history must be recognized as gendered.

Candlelight, Women, and the Material Culture of the Everyday

During the course of the nineteenth century, just as in the century before and as we see described in the epigraphs heading this chapter, housewives and domestic servants continued to have their work (paid and unpaid) framed by the need for artificial light.[31] This was gendered in two ways: either by role, or by the quantity of time spent on a task. Many commonplace, domestic routines undertaken by women depended on illumination, even if that was light spilling over from the fire. For this reason, there were glass magnifiers designed to throw candlelight onto close work such as darning and sewing. Nursing and childcare, work that extended women's domestic working day, also often required light late at night. The use of lighting and specific equipment designed to facilitate that use could therefore be shaped by gender-specific roles, such as equipment for (ongoing) use in the nursery, which was designed to cast a gentle light and could also be used to heat food for a baby or for someone requiring care.[32]

From the evidence of material culture, we know that, thanks to their adaptability, there were some persistent domestic uses for candles that fell within the gendered bounds of caregiving, childrearing, and everyday nursing.[33] The management of lighting equipment was also laborious in and of itself; it took physical effort and skill, for example,

2.1a and 2.1b   Food warmer, metal: base height, 18 cm; diameter, 11.5 cm; tray height, 8 cm; tray diameter, 9.5 cm. Candle cage: height, 6.5 cm; diameter, 9 cm; lid diameter, 9 cm. This particular candle cage came from Adderbury in Oxfordshire, where it was last used in a nursery for a baby who later lived for eighty years and died in the 1920s.

to clean ormalu candlesticks and their branches and lamps "with soap and water," to use "hot water and pearl ash" to clean oil lamps that were "foul inside," and to use "soap and water" to get rid of "burnt spots" from the lamp glasses. Although in the wealthiest homes these tasks, and jobs such as regularly replacing candles in candelabra or oil in lamps, might be the province of male servants, they were more frequently taken on by women. Servants were normally referred to in advisory texts as "she" or "her." In this respect this work was, societally, gender-neutral, but in practice it used more of women's time.[34] In both cases this was valuable work undertaken for the household – an exercise of production and of resource management and control – by women.

Finding Women's Lives Lit by Candlelight in the Archives

The interaction of women and young children within and around candlelight is revealed through a case of theft in 1830, in which the defendant, a woman who lived in a tenement, was said to have told a (female) witness to "take up the baby, light a candle, and play with him."[35] It is not only in incidental evidence from court cases capturing the life of the urban poor that we see candles. There are also traces of them in country life, in the periodical press aimed at the well-to-do. According to the *Farmer's Magazine*, in the 1830s and 1840s, candles were used to light bedrooms – for getting up and getting dressed before daybreak to catch an early coach, or by students for reading at night – while managing the known risks carefully.[36]

Candles were ubiquitous across the social classes and in all walks of life in the 1830s. As the Victorian period ended, despite the widespread use of oil lamps (revivified by kerosene) and the adoption of gas, this was still true. Advisory, prescriptive texts such as *Mrs Beeton's* outline the gendered (and thrifty) bounds of energy use very well, setting out for the aspiring bourgeoisie what a servant (the housemaid) needed to do to keep an affluent house well-lit, the work entailed in managing light demarcated by social class. In the 1907 edition, for instance, it was stated that lamps, gas and electric globes, candle sticks, etc., have to be cleaned, "and here the housemaid's utmost care is required. In cleaning candlesticks … she should have cloths and brushes kept for that purpose alone; the knife used to scrape them should be applied to no other purpose; the tallow-grease should be thrown into a box for that purpose." Instructions were given to the employer regarding managing the ready availability of light at night, via the work of their servant: "And now … everything in the room in order for the night, the housemaid [must take] care to leave the night-candle and matches together in a convenient place, should they be required."[37] Each edition therefore reveals incidentally which energies were likely to be used in a well-to-do home, the (typical) continued use of candles throughout this period, and the transactions of work across the boundaries of gender and class. In *Enquire Within upon Everything*, a publication aimed at the aspiring upper working class or aspiring lower middle class, which was revised frequently "in accordance with the progress of the times," we find a series of tips, this time aimed at the mistress of the house, in the section called "Hints for Home Comfort." On reading by candlelight in bed, readers are told: "have the candle behind you"; on sewing by candlelight: "sew against a white cloth/sheet of paper"; and on avoiding both candle-grease spots and "risks of fire": "do as much housework as possible before nightfall in winter, to avoid running around with lit candles."[38] Candlelight was evidently entangled with leisure (reading in bed), and had to be managed carefully (with the use of white paper to maximize the available energy, for thrift or practical reasons), and carried risks that might restrict a woman's working day.

Candles remained so commonplace even when other forms of lighting were possible that their availability was taken for granted, but sometimes their production and use could be determined structurally by women's domestic roles. Though the 1898 edition of the *Sunlight Year Book* carried advice on reading a gas meter, for instance, it also outlined the following treatment for earache alongside an illustration of a nursemaid: "Warm a small spoonful of glycerin (Lever's) by holding it in a spoon over a candle, pour gently into the ear."[39] A candle was flexible – it could be moved to where its light (or its heat) was needed, and it could be used on the move. Though by this point most people acquired candles commercially as opposed to making their own rushlights or "dips" at home,[40] candles also dripped and left stubs, and this waste material could be reworked. Hence, in the 1894 *Enquire Within upon Everything* (four years prior to the 1898 *Sunlight Year Book*) though the publication recommended buying Field and Childs's nightlights,[41] it also offered instructions on a homemade version of these using cotton and candle waste, set in pill boxes and placed in a saucer of water, as in "this manner, the ends and drippings of candles may be used up."[42]

Candlelight and Cost

Those who trained servants in the science of the domestic economy sought to wean the household off commonplace candlelight, because, according to William Tegetmeier in his *Manual of Domestic Economy* (1880), for example, they were deceptively expensive, which explains the tips on reduced use and recycling, also directed at women, which were framed as an essential element of household management. "Candles, because of the cheapness of their first cost," Tegetmeier stated, "are largely employed as a source of artificial light; but from the small amount of light yielded by them, they are almost the dearest source of light we employ."[43] Determining the exact expense in candles as compared to other light sources is difficult. However, seventy years earlier, in Accum's *Practical Treatise on Gas-light* (1815), we see that this issue was clearly important enough to readers deciding to adopt a new form of energy that he listed several sets of workings, comparing individual cases of off-grid gas manufacture and use compared to the costs of using commercially sourced tallow candles.

Accum calculated that 2,500 "mould candles" at one shilling per pound, using four-tenths of an ounce of tallow per hour at an average daily use of two hours, would cost around £2,000 a year, compared to gas at £600, to generate the same amount of light in his cotton mill in Manchester. A similar mill burning candles for an average of three hours a day would incur £3,000 in expenses in candles, compared to £650 for gas. A smaller metal toy factory in Birmingham used eighteen to twenty-four gas lights in winter at a cost of four pennies per day, compared to three shillings a day for candles, that is, £50 a year based on twenty weeks of use. Another small-scale thimble manufacturer in London said that it cost them seven shillings to burn forty-two tallow candles for seven hours (seven pounds candles at one shilling per pound). In all cases, this was based on the generation of gas on site. Accum was seeking to promote the adoption of gaslight (and it is beyond the scope of this qualitative study to test the figures, so they must be taken cautiously), but as he was also seeking to convince his readership, we may reasonably read this as indicative of the costs of using either moulded tallow candles or off-grid coal gas at that time. This also accords with Fouquet and Pearson's argument that the well-to-do looked for and were better able to afford expensive efficiency gains in the adoption of new lighting technology.[44] Nevertheless, in Tegetmeier's statement (within a chapter that goes on to discuss oil light and gaslight), and in the expectation in the 1907 edition of *Mrs Beeton's* that candles ought to be within easy reach at night (when electricity was already available to the urban bourgeoisie), we have clear evidence of candles persisting as part of a mix of domestic energies, even in a better-off household during the early years of the twentieth century. Candles were therefore in use in Victorian households (and elsewhere) for the whole of this period, and both servants and mistresses were expected to work with them.

In describing the wife or the housemaid's allotted tasks, *Enquire Within upon Everything* outlines the necessary gendered work involved for the good management and use of lighting at home. But, as noted, there was also class in candles. In *Mrs Beeton's*, class can be seen in the paid tasks – scraping candlesticks and catching the pieces, the careful placing of candles and matches when setting up a room for the night – that have been shifted out of the reach of the middle-class members of the household. In other sources we find class playing a role in the consumption of particular types of candle. Setting aside the quantitative aspects of cost and efficiency, energy and materials were equated with social standing, because of the necessary purchasing power required. In literature, we find that everyday assumptions about what was thought to be required by way of lighting varied by class and by gender. In *The Tenant of Wildfell Hall* (1848), a man is considered inebriated, uncouth, and rude if he could not tell that his hostess's candles were "wax; they don't require snuffing"[45] – it is reasonable to suggest that its author, Anne Brontë, would only have used such a statement if she knew that her literate female readers would understand the social and class significance of the wax candle, and the management of light. We see something similar in the witty description of saving the cost of expensive candles at social occasions in Elizabeth Gaskell's *Cranford*.[46]

We also read of a woman who reportedly stole candles because she wanted to give the appearance of having a higher station than she actually had. According to a letter in the *Daily News* (1856), the woman was said to be separated from her husband, a "pianoforte tuner ... and lately keeping a lodging-house ... [and] there is no doubt that her efforts to make an appearance in society above her honest means prompted her to commit acts for one of which she is now undergoing a just but no doubt severe punishment."[47] The prosecutor, the tradesman from whom she stole, had six other businesses in London and wished to protect his reputation. He therefore wrote a defence of his case to the paper – in which an earlier letter defending her had been published by one of her lodgers. In the tradesman's description of the case, in court, he claims to have seen her enter his shop while he was getting some oil, and observed her going to "the pile of candles" instead of the counter. When she spoke with him about something that she was buying, he "perfectly and distinctly saw the end of the packet of candles under her arm." When asked, she said that she did not want to buy anything else, and went towards the door rather than to the pile of candles. He stopped her at the door and lifted her mantle to reveal the packet of candles. She then said that she wanted the tradesman "to send them home" ("Do send them home, there's a dear creature"). To which he replied, "I shall not; but shall send them along with you to the station-house." This interaction was observed "by the wife of a professional gentleman."[48] From this case, it is clear that for bourgeois Victorian women, decision-making about the maintenance and use of particular sources of energy for lighting was closely related to social standing and respectability, something that was a mix of social class and gendered identities, and not just a matter of the quality or efficiency of illumination.

Candlelight and Labour

The following description of tallow candles, and their comparison with wax candles, appears in Accum's *A Practical Treatise on Gas-Light*:

> The flame of a tallow candle will of course be yellow, smoky, and obscure, except for a short time after snuffing. When a candle with a thick wick is first lighted, and the wick snuffed short, the

flame is perfect and luminous ... As the wick becomes longer ... the tallow ... is less completely burned, and passes off partly in smoke. This evil increases, until at length ... a portion of coal or soot is deposited on the upper part of the wick, which ... at length assumes the appearance of a fungus. The candle then does not give more than one tenth of the light which the due combustion of its materials would produce; and, on this account, tallow candles require continual snuffing. But if we direct our attention to a wax candle ... the wick being thin and flexible ... when its length is too great ... [it] bends on one side ... Hence it ... performs the operation of snuffing itself, in a much more accurate manner than can ever be performed mechanically ... [T]he important object to society of rendering tallow candles equal to those of wax ... depend[s] on ... the degree of fusibility in the wax: and that, in order to obtain this valuable object ... is, in a commercial point of view, entitled to assiduous and extensive investigation.[49]

Because certain materials resulted in more (or in less) human effort in managing the best effect in terms of illumination – that is, human labour – the cost or price of a candle could be matched to the placement of energy either within the material burned, or in the work required to trim or snuff a candle. Though this knowledge was well-understood by science, and disseminated by authors like Accum from the early nineteenth century, it was reiterated in advisory literature written to educate the servant class.[50] Texts like Tegetmeier's judiciously set out the science for their target readers: girls in training institutions and girls' schools.[51] The selection of the correct energy for the generation of artificial light was therefore, in his view, an expression of careful "domestic economy" – a view that was evidently influential, as the book had gone into ten editions by 1880, and was referenced in a government paper on education. This type of economic knowledge was seen as useful enough to help girls obtain better-paying jobs.[52]

At the time, the price of tallow candles remained stable relative to other costs. Wax candles always cost about three or four times as much as tallow,[53] but the cost of all candles was a perennial issue, and in 1839 the *Times* included a tip about making some savings, even on tallow: "use salt to slow down the burning of the tallow, that will help a candle last all night with enough light for a bedchamber."[54] Meanwhile, oil costs were kept high by government intervention, and by low competition at sea.[55] It therefore appears that a sufficiently high enough degree of capital and commercialization was implicated in the consumption of artificial domestic light to lead to an encouragement of household thrift. As well,

2.2   This candle reading lamp is nickel plated. It stands 12.5 inches (32 cm) closed and 15.5 inches (40 cm) extended. The diameter of the base is 5 inches (12 cm).

before the market for gas began, the affluent householder was already dependent on the more highly processed of the available fuels, because they had shifted their purchasing power over the material energy of the wax candle (or lamp) through the selection of a higher-quality product that required no labour at home in the making of illumination, and less work in its management. Those who had additional money at their disposal could also, by the end of the period, purchase labour-saving holders designed to move a wax candle gently upwards as it burned, thereby maintaining the exact position of the light for reading without intervention, either by the owner of the candle or by their servant.[56]

Conclusion

While teleological, progressivist historical accounts using official records and datasets, and patents and their associated legal cases, have split the history of illumination into bite-sized chronological moments of dramatic technological change, if we factor in the history of home life and the rather more ephemeral sources associated with those lives, such as advisory literature and household manuals that are more closely aligned to and representative of women's lives, then we can see the continuities in the history of lighting and therefore also of energy. Though not the subject of this chapter, which concentrates on use, not production, it must also be remembered that candles (and other lights) depended materially on an ongoing interconnectedness of animal, vegetable, and mineral "resources" and energy flows within the Victorian nation, empire, and global marketplace. As the use of candles increased in the 1830s, for example, so more wax, and tallow (beef and mutton fat, spermaceti, and vegetable oil), was sourced and sold. Tallow sales and prices in their turn affected graziers, and manufacturers of other products, such as soap.

The candles made and bought in nineteenth-century Britain, like food and other resources used within the global North, depended on raw materials from "ghost acres" and agro-ecological environments, and on human and other animal labour far beyond Britain's shores. Hence, the *Times* reported a heated debate on the duties due on soap and candles in January 1836, when H. Handley, Esq., MP, argued that the three shilling, four penny per hundred weight of duty on tallow imports (at a rate of 50,000) from Russia should be raised to ten shillings, or 10 pounds, per ton. Annual production of tallow in UK at the time was reckoned to be 100,000 tons. Subsequently, soap manufacturers called for tax to be taken off soap, as it had been from candles, alkali, and glass. Without taxation, and therefore without tax records, continuing through the period for domestic tallow production, it is not possible to reliably assess the quantity of tallow produced from UK livestock. But with large quantities of tallow in various forms coming in through London and Liverpool from Russia, it is hardly surprising to see that these raw materials figured large in the very heated political debates of the period regarding protection versus free trade, and the competing commercial interest of different industries, or that these debates were considered newsworthy.[57]

Yet, by seeking out women within the history of artificial lighting, we find a different complexity within the history of energy, in this case the lasting history of energies used to generate domestic illumination and, sometimes, small amounts of heat. We see the potential energy captured in candles – already made up of multiple energy transfers (including the work of production) – not just being released, but being laboriously redirected by yet more work, such as snuffing the candle to produce light (not smoke) and a little heat to warm food at night for an infant or for someone being cared for, in continuous and necessary acts of nurture and care. We see unused energy saved, and yet more energy added through work as dripped wax was scraped off, then refashioned with candle stubs into yet more candles, ready to light. We find, too, at the point of use, the well-to-do using money to shift energy consumption away from human labour and employment (snuffing and trimming tallow candles), and towards the consumption of condensed chemical energy (self-snuffing wax and patent candles). We see the meanings attributed to candles inflected by class, as well as by gender; we see candles shaped by homely utility, and by emerging competition for new products. And, in recognizing within this the continuing domestic demand for candles – portable, and easily adapted to late-night care, not just illumination – we find that decision-making about energy use in the nineteenth century belonged to the "woman's sphere."

Notes

1   The history of artificial lighting in Britain has been told and retold by, among others, O'Dea, *The Social History of Lighting* (1958); Schivelbusch, *Disenchanted Night*; and Bowers, *Lengthening the Day*. For a nineteenth-century account, see Accum, *A Practical Treatise on Gas-Light*, or for a geographically wider-ranging, longue durée treatise, see d'Allemagne, *Histoire du Luminaire*, cited by O'Dea in his acknowledgements.

2   See Robert Browning's poem "Meeting at Night" (1845), and Briggs, *Victorian Things*, 45, 50, 69, 179.

3   Nead, *Victorian Babylon*, 92–3.

4   Fouquet and Pearson, "Seven Centuries of Energy Services," 139–77. For a recent account of vegetable fat, and "ghost acres," see Robins, "Oil Boom," 313.

5   O'Dea took ten years to write *The Social History of Lighting*, based on the collection. He outlined the various substances used to fuel light and to make candles (213–20).

6   "Multiple Classified Advertising Items," *Illustrated London News*, issue 86 (23 December 1843): 415. In this publication, an advertisement from Price's is followed immediately by one from W.S. Hale. The Price's advertisement explains both labelling and patent symbols, and the need for these due to imitators. The advertisement stresses that Price's does not use arsenic. It provides prices for various sizes, weights, and materials, pointing out those that require snuffing and those that do not, and the brilliance and purpose for each product (for example, their "Composite Candles" require no snuffing, and cost one shilling [1s] per pound, whereas their "Coco Nut Candles" cost 10 pennies [10d] per pound). W.S. Hale focuses on their

experience in making composite candles, and also stresses that there is no arsenic or other "deleterious materials" in their candles.

7   "Fire – Destruction of Messrs. [ill] Candle Manufactory," *Derby Mercury*, 25 November 1846.

8   *Times*, 30 June 1837, 6, col. B; *Times* 9 November 1837, 3, col. E; Bowers, *Lengthening the Day*, 20.

9   *Times*, 14 November 1873, 10, col. F.

10 *Times*, 7 March 1889, 3, col. A.

11 *Times*, 9 November 1837, and 14 November 1873.

12 See figures regarding "Price's Patent Candle Company Limited," *Liverpool Mercury*, 22 March 1884.

13 *Times*, 23 January 1873, 10, col. F. The company's candle exports valued £132,658 in 1870, £180,548 in 1871, and £223,452 in 1872.

14 Each form of illumination carried varying social and cultural meanings, often linked to notions of empire. See Sayer, "Atkinson Grimshaw, *Reflections on the Thames* (1880)," 17–19.

15 See Accum, *A Practical Treatise on Gas-Light*; as well as Alglave and Boulard, *La lumière électrique*.

16 Case studies can be found in Otter, *The Victorian Eye*; and Gooday, *Domesticating Electricity*. Cultural histories for the period and location – on the meaning and use of illumination, the production of the material goods associated with artificial light, the sources of ignition, and the built spaces of energy generation – include Briggs, *Victorian Things*; and Nead, *Victorian Babylon*.

17 Falkus, "The Early Development of the British Gas Industry," 218–19.

18 Schivelbusch, *Disenchanted Night*, 28.

19 Nead, *Victorian Babylon*, 103.

20 Ibid., 103.

21 Schivelbusch, *Disenchanted Night*, 29.

22 Nead, *Victorian Babylon*, 103.

23 For an example of a secondary source returning value to women's household work, based on historical sources see, Romines, *The Home Plot*.

24 This text is followed by the footer: "sunlight soap makes homes brighter and hearts lighter." *Sunlight Year Book*, 301–2.

25 *Sunlight Year Book*, 1898, 382, 383.

26 Hall, *White, Male and Middle Class*, 1–2.

27 Kiechle, *Smell Detectives*, 78–80, 146–7.

28 Scharff, "Introduction" in *Seeing Nature through Gender*, xiii. Her exact statement is, "Gender, the bundle of habits and expectations and behaviours that organizes people and things according to ideas about the consequences of sexed bodies, is a crucial, deep, and far-reaching medium through which we encounter nature. Gender varies from time to time and place to place.

But all humans, in all cultures, think in and act in gendered terms ... In
short, humans know nature, at least in part, through gender."

29 Haraway, "Anthropocene, Capitalocene, Plantationocene, Chthulucene."

30 Stephenson, "Sustainability Cultures and Energy Research."

31 Collier, "The Woman's Labour." Duck's poem was originally published in 1730.

32 Florence Nightingale gained her reputation as "the lady with the lamp"
    because of the religious connotations of the phrase (akin to Jesus as "the
    Light of the World"), but also because she needed a portable soft light
    to oversee the wounded on the wards. See Pollard, *Florence Nightingale*,
    frontispiece, and 99.

33 Schwartz Cowan, *More Work for Mother*, xv–xvi.

34 Employment of a male servant was indicative of a high social standing,
    because of the expense, and was taxed as a result. Most middle-class and
    lower-middle-class households employed one to two female servants;
    those who had more would take on a cook (female), a chamber maid or
    housemaid to clean and wait at tables and answer the door, and a scullery
    maid, but only the better off (upper-middle-class and above) households
    took on a manservant. For one of the earliest histories of service, see
    Marshall, *The English Domestic Servant in History*. But historical work on
    service really developed substantively mid- to late twentieth century with the
    emergence of women's history, such as McBride, *The Domestic Revolution*; and
    Sharpe, *Adapting to Capitalism*. For advice to "housewives in the middle ranks
    of society" regarding "hiring domestic servants" and instructions for cleaning,
    see *Chambers's Information for the People*, 647–9. For a discussion of gendered
    and gender-neutral domestic tasks that engaged women in the home more
    than men, see Schwartz Cowan, *More Work for Mother*.

35 *Old Bailey Proceedings Online*, February 1830, trial of Elizabeth Hallick.

36 They also lit work outside around larger houses – for example, they helped
    grooms to care for horses. For examples of these various typical uses, see *The
    Farmer's Magazine* series (1835), 350 and 272; (1845), 517; and "Precautions as
    to Fire," *Chambers's Information for the People*, vol. 2, 644.

37 Beeton, *Mrs. Beeton's Household Management*, 1780. The text continues: "the
    same with everything connected with the lamp-trimming; always bearing in
    mind that without perfect cleanliness, which involves occasional scalding, no
    lamp can be kept in order."

38 *Enquire Within upon Everything*, preface, iv, and paras. 24, 26, 36, and 104.

39 *The Sunlight Year Book*, 1898, 354, 380.

40 The making of these by female relatives and their use was recollected by
    William Cobbett (1763–1835) and George Baldry (1864–1955). Cobbett
    wrote, "My grandmother ... used to get the meadow rushes ... cut them ...
    take off [most of] the green skin ... [and] put [the rushes into melted]

grease"; and "I was bred and brought up mostly by rush-light ... to sit by, to work by, or to go to bed by, they are fixed in stands made for the purpose ... The stands have an iron port something like a pair of pliers to hold the rush in, and the rush is shifted forward from time to time, as it burns down." Cobbett, *Cottage Economy*, 144–5. Baldry said his mother always resisted using anything other than rush-lights, as she was wary of new ways or extra expense. Baldry, *The Rabbit Skin Cap*, 37, 208.

41 Field and Childs were competitors to Clarke's nightlights and decorative "fairy lights."

42 *Enquire Within upon Everything*, para. 1001, 158.

43 Tegetmeier, *A Manual of Domestic Economy*, 39.

44 Ibid., 38; Accum, *A Practical Treatise on Gas-light*, 66–76; Fouquet and Pearson, "Seven Centuries of Energy Services," 174–5.

45 Brontë, *The Tenant of Wildfell Hall*, 275.

46 Gaskell, *Cranford*, 112–13, 118.

47 "The Case of Candle-Stealing by a Lady," *Daily News*, 4 October 1856. "Letter to the Editor" of the *Daily News* (by "the prosecutor in the case," i.e., the tradesman from whom she stole).

48 Ibid. The prisoner was said to have done something similar at a grocer's shop, taking a pot of marmalade and, when accosted, saying she wanted it added to her order. In that case, the shop owner "thought she was a respectable person [so] he accepted her explanation, and charged the pot of marmalade in the bill."

49 Accum, *A Practical Treatise on Gas-Light*, 18–21.

50 Accum cites "Mr Nicholson, 'Philosophical Journal,' 4th series, vol. 1, 70." *A Practical Treatise on Gas-Light*, 21.

51 Tegetmeier, *A Manual of Domestic Economy*, 37. Tegetmeier explains, by way of analogy: "The fat is first melted, ... [at] the centre of the flame ... it is, by the high temperature, converted into gas, which, burning in the air, produces a flame, so that a candle is in reality a portable gas-lamp, manufacturing the gas as it is consumed."

52 Ibid., preface.

53 Falkus, "The Early Development of the British Gas Industry," 217–19.

54 *Times*, 4 October 1839, 3, col. F.

55 Here we see the linking of light consumption with much wider issues. The British whaling fleet – in 1816 Britain sent 150 ships to the territory covering Greenland, Baffin Bay, Hudson Bay, Davis Strait, and the North Atlantic – was seen as a training ground for sailors, and as demand for whale oil increased, the government encouraged its trade. In 1750 it doubled the bounty (from twenty shillings per ton per voyage to the north, to forty shillings); during the American Revolution the Royal Navy harassed American whalers; then,

postwar, the government protected the British industry with high tariffs. At
the end of the French wars there was apparently less need to train sailors, so
the tonnage bounty was lost in 1824. Davis, Gallman, and Hutchins, "The
Decline of US Whaling," 569–70, 584–6; Bowers, *Lengthening the Day*, 20.

56 See the candle reading lamp, manufactured by Falk, Stadelmann & Co. of
    Greater London in the nineteenth century. It is nickel plated, and has the
    Falk number P42617. MERL Object number 57/398/1-2.

57 *Times* 5 January 1836, 3 col. A.; *Times* 5 January 1836, 3 col. A.; *Times*
    1 February 1851, 3, col. F.; Inspector-General of Imports and Exports,
    e.g. "Mr Collier – Return of the Quantities of Linseed, Corn, Hemp, Flax, or
    Tallow Imported during the Year 1854, into the United Kingdom, from the
    Black Sea…," 9 March 1855 (MERL Pamphlet 4822 Box 1/11). The history
    of the British Tallow trade, i.e. imports as recorded by the Board of Trade
    during the nineteenth century, has been discussed recently by Jim Clifford re
    the soap trade. See Clifford "London's Soap Industry"; Fouquet and Pearson,
    "Seven Centuries of Energy Services"; and Robins, "Oil Boom."

3

Women, Fear, and Fossil Fuels

R.W. Sandwell

Yesterday evening Mrs. Wm Fisbell ... attempted to light a fire
with kerosene. Results – House consumed, child burned to death,
Mrs. Fishell and another child probably fatally injured.

London Advertiser [Ontario], 3 August 1880

As climate change, intensifying fossil fuel pollution, and the COVID-19
pandemic are providing significant challenges to familiar triumphalist
energy modernization narratives, it seems like a propitious moment in
history to revisit the apprehensions that often framed the experiences
of women encountering new energies in their homes for the first time
in the nineteenth and early twentieth centuries. Historians exploring
women's experiences of changing energy use in their homes have
tended to emphasize women's rational decision-making about cost
and performance, and the cultural, economic, and patriarchal contexts
of a consumer society that set the stage for the changes they embraced,
or that were thrust upon them.[1] Women's energy anxiety can be a
difficult topic for feminist historians to broach, its effects impossible
to confirm, and the evidence of it usually far from conclusive, when it
can be inferred at all. There is little statistical evidence documenting
the actual extent of bodily harm resulting from these new energy forms
in homes of this era, nor is there any way of reliably measuring the
extent to which fear and anxiety influenced the typically slow uptake
of new energy carriers.[2] Anxieties clearly did not prevent women from
buying and using new kinds of energy. But as Graeme Gooday argued
so persuasively in *Domesticating Electricity*, fear and anxiety lurk just
beneath the surface of women's energy narratives, informing their energy
decisions and behaviour, shaping the ways that appliance and utility
companies sold their products and services, and contributing to the
rapid growth of a vast, amorphous educational empire of advice about
household management.[3] Women were well aware that the new fuels
were not only more powerful, but considerably more dangerous than

the ones they replaced. Their fears were at their most acute when the new fuels were least familiar.

Notwithstanding a gendered discourse that often trivialized women's timidity and criticized their ignorance in the face of new energy carriers, many women continued to engage with them on a daily basis, at the same time trying to protect their families and mitigate the harm that new household fuels could inflict. For in the 1850 to 1930 period, before rigourous safety standards were in effect, the onus was firmly placed on women themselves to negotiate safe use.[4] And they worried about it. Drawing on a range of British, American, and Canadian sources, opinions, and discussions with which many Canadian women would have been familiar by the late nineteenth and twentieth centuries, this chapter explores women's early encounters with new fuels in their homes. Some of these encounters took place on actual geographical frontiers of settlement in Canada. For other women, the frontier was metaphorical, comprising their everyday domestic encounters at the interface of modern energy use.[5] This chapter focuses on three key sources of women's particular energy anxiety in the pioneering era of fossil fuels: the fires, the explosions, and the toxic air identified specifically with gas and oil.[6] This chapter, then, is an exploration of the darker side of the transition to fossil fuels.[7]

The Home as a Space of Danger

Historians on both sides of the North Atlantic have explored in some detail the emergence in the nineteenth and early twentieth centuries of domestic space as a "haven in a heartless world," a gendered, idealized interior world where a woman could isolate and protect her family from a hostile outside world of commerce, business, pollution, and the other unpleasant material and cultural consequences of rapid industrialization and urbanization.[8] Historians have long challenged the nineteenth-century belief that women's domestically focused lives actually existed within a physically defined sphere independent of and materially separated from the public, outside male world. But there is general agreement that women of this era became increasingly tasked with maintaining the happiness, health, and well-being of family members in the spaces of the home.[9]

As such, the home was also becoming a place of increasing anxiety for and about the women held responsible for the ever-expanding amounts of work that the highly gendered project of home health and well-being demanded. There was a lot to be concerned about. By the mid-nineteenth century, a burgeoning "santitarian" movement was spreading across North America and Europe.[10] Doctors, sanitary engineers, plumbers, architects, and social reformers had identified "house disease" as a potent source of danger to individuals, families, and society at large: poor sanitation, poor hygiene, and communicable diseases had to be constantly monitored in the home, for "damp cellars, dusty carpets and dank water closets were domestic breeding grounds for the invisible agents of deadly disease."[11] By the late nineteenth century, women (and primarily in their role as mothers) had been identified as "home safety managers"[12] whose paramount job had become "to protect loved ones from disability and death,"[13] guarding not only against disease, but against hazard as well.

Historians have documented how a wide range of educational initiatives, from government public health directives, anti-tuberculosis campaigns, and women's institutes and homemakers' clubs, to the efflorescence of homemaking manuals and domestic science and home economics educational materials, all sought to assist homemakers, present and future, in their monumental, and indeed heroic, task of caring for their families.[14] By the late nineteenth century, women's role as home safety and family well-being managers had become entangled in an ever-widening range of issues that found expression in the burgeoning home economics and domestic science movements.[15] A significant component of those movements of education and social reform was the re-education of women for the modern energy regime: for the first time, the ways that women provided heat, light, and food, and how they carried out a variety of home maintenance tasks, became the subject of considerable public and indeed corporate interest across Europe and neo-Europes by the early twentieth century.[16] Women's energy use in the home was deeply embedded in the energy triumphalism that dominated discussions of energy transitions in the nineteenth and twentieth centuries.[17] But fossil fuels also created considerable anxiety for women who were using them for the first time in their homes.

Understanding Energy Anxiety

To homemakers, the new forms of energy they were being actively persuaded to welcome into their homes often seemed expensive, extravagant, and, moreover, untested. As Gooday has detailed in *Domesticating Electricity*, it required lot of work to turn "a mere technological possibility into an actual household experience"; women had to be convinced that electricity had been "sufficiently tamed to be safely, reliably and comfortably introduced into the home."[18] The relative cost, convenience, and utility of new forms of energy were uneasily weighed against the increased leisure, convenience, comfort, and status promised by the promoters of the new energy sources, and particularly in the early years when gas and electricity were new and unfamiliar.

Women's energy fears and anxieties emerge erratically in the historical record. Notwithstanding the heightened interest in the home from "outsiders" such as doctors, plumbers, electricians, engineers, and a variety of reformers, there is very limited statistical data documenting the nature or extent of home hazards before the second half of the twentieth century.[19] Women's energy fears were seldom legitimized in engineering tracts, appliance advertising and promotion, newspapers, or home advice manuals, but, as we will see, many of these sources articulate for today's readers, often with startling clarity, exactly why women had cause for concern. While companies routinely dismissed dangers relating to their own energy carriers, such as electricity or gas, they were quick to demonize competing ones, providing direct evidence of safety concerns.

Fear and anxiety can also be perceived in the inevitable advice that household and other manuals offered to women about their appropriate relationship to various appliances. The tone and content varied. Some attempted to assuage women's fears: "Kerosene is not explosive. A lighted taper may be thrust into it, or flame applied in any way, and it does not explode."[20] Others provided detailed explanations of how women might mitigate the danger and overcome their fear: "When you know how to protect yourself and are sure that no matter what happens you are able to avoid injury you will have confidence."[21] Advice manuals and newspapers often scolded women for their unreasonable timidity, carelessness, and ignorance: "The feminine propensity to kerosene suicide seems incurable. In Brooklyn yesterday a servant new to the country and fresh from warnings not to use oil in making her fires repeated the old experiment with the old result."[22] Indeed, a persistent theme was women's ignorance and backwardness in failing to adopt new kinds of energy, or doing it badly, part of a broader critique of women's much-lamented backward technological abilities in a modernizing world. As one home economist summed up, "It has been said that housekeeping is the most backward science in the world. It was no doubt a man that said if women went away for six months and left the men to do the work, they would on their return find the men sitting on the fences smoking while machinery did the work."[23] Aware that their fears were often deemed unreasonable in the eyes of men, women nevertheless also expressed them directly in diaries, letters, and oral histories. These sources document that women's fears were focused on a wide variety of perceived dangers about new forms of energy in their homes. Three of the most persistent were fire, explosions, and air quality.

Fire

Fire was always a prime source of concern within the home, and particularly when all cooking, lighting, and heating was done by means of a more or less open flame. Lighting was a gendered activity in the nineteenth and twentieth centuries, and one of particular concern as a source of accidental fires. When kerosene began to replace the older fish, animal and vegetable oil in lamps, women continued the traditional work of trimming, cleaning, and tending lamps and candles, which Karen Sayer has documented in this volume.[24] The dangers of moving a candle or lamp from place to place as the demand for light required were well-known to nineteenth and twentieth century households. Children were taught from an early age never to run with a lighted flame. Candles in sleeping areas, particularly children's bedrooms, were generally prohibited, except in times of illness.[25]

The dangers of lighting were widely discussed. As Robert Hammond, of Hammond Light and Power Company, pointed out in his fiercely partisan defence of electric lighting in 1884, "fires … may almost always be attributed to something connected with lighting." They were, he declared, largely due to carelessness:

> a candle is left burning by the bedside and sets the bedclothes
> on fire; a naked gas-bracket is pushed too close to the wall
> and the woodwork of the room ignites, a lamp is accidentally
> overturned … and the room is in a blaze immediately, a falling
> candle touches an actress's dress and a life passes away in agony;
> a spark from the taper that is used for lighting the gas smolders
> between the oaken boards … the ignition of the escaped gas from
> a gas jet left on all night, in a small room, causes an explosion
> and fire.[26]

It was kerosene lamps, however, that were identified in the later–nineteenth-century press and advice manuals as a particular fire hazard. Although lamp manufacturers and petroleum salesmen argued the increased safety of the coal-oil lamp over candles and older lamps because the flame was contained by a glass chimney, this petroleum product was (and is – it is now used primarily as a jet fuel) significantly more flammable than the organic illuminating oils it replaced, with the notable exception of dangerously explosive camphene (an organic derivative of trees via turpentine).[27] Even after standardization in the refining and testing of kerosene was credited with reducing the number of kerosene-ignited fires in the late 1870s, fires resulting from "upsetting or bursting of a coal oil lamp" appear with regularity in the newspapers.[28] In Canada, where kerosene lamps continued to dominate artificial lighting for the country's substantial rural populations until the Second World War, kerosene's relationship to fire was frequently discussed.[29]

While no statistics on kerosene accidents were gathered for the United States or Canada as a whole, Cleveland (where Standard Oil was headquartered at the time) documented an increasing number of fires from kerosene, from 56 in 1865 to 249 in 1874. Eleven fires were reportedly from exploding oil, five from spontaneous combustion, and thirteen from exploding kerosene lamps.[30] A 1902 British educational treatise on the use and care of the petroleum lamp, while providing more statistical information about the dangers, downplayed what newspapers sensationalized as "domestic dynamite." Tragic as the consequences of kerosene fires were, the authors pointed out that despite the dangers of the coal-oil lamp, it still compared favorably to the incidents of death from falling downstairs (731 per annum), or to the 1,684 children over a two-year period "burnt to death through being left alone in a room with a fire."[31] The author suggested that most of the 134 English deaths on average per annum from petroleum lamps in the last decade of the nineteenth century could be traced to carelessness in moving the light, or in placing the lamp on an unstable and flammable surface, where the flame might catch spilt oil. The Ontario Monetary Times noted in 1895 that there had been 1,598 fires reported that year, with 360 "put down to preventable causes" such as "sparks, matches, defective flues, defective chimneys, lamps and lanterns, stoves and pipes, spontaneous combustion etc."[32]

As noted, the carelessness, ignorance, and general culpability of women, and particularly servants (almost invariably identified as women), in causing fires was a common theme. In 1880, Scientific American reported that "The principal causes of fires have been carelessness on the part of servants or occupants of houses (this is accountable for nearly one quarter of all the fires) ... [caused by] foul chimneys, explosion of kerosene lamps, and window curtains near gas jets."[33] In 1923, the American Underwriters Association launched an issue on home fire prevention for the first time, insisting that most fires were preventable; they "blamed inattention and carelessness for 75 per cent of the 15,000 annual fire-related deaths, half of which were in the home."[34] Speakers at the National Safety Council's Safety Congresses in the same year had no doubt where the responsibility for fire prevention in the home was located: it was "women's responsibility to prevent carelessness with matches, to avoid the use of inflammable cleaning fluids, to stop the misuse of kerosene lamps, gas stoves and flatirons, to avoid the piling up of oily rags and clothing, and to be vigilant for excessive wear of electrical wires."[35]

Women had also been responsible for managing the other major source of open flame in the home: the cooking, lighting, and heating fires of the open hearth. Women had always been well aware of the necessity of keeping garments and long hair – as well as small children – away from the open flame as they tended and stoked the fire, and in moving pots, pans, and spits nearer to and further from the source of heat. When the use of new cast iron coal and wood stoves became widespread in North America in the 1860s, manufacturers were quick to identify their enclosed firebox as an important safety advance over the open hearth. But cracks, poorly fitting plates, and unstable legs in cheaper metal stoves meant that dangerous sparks as well as smoke continued to pose dangers from fire related to cooking and heating.[36]

The stoves also introduced new dangers. Stovepipe fires became an increasingly well-known hazard, as burning wood in an enclosed stove creates a highly flammable residue of creosote. This was impossible to manually remove from the pipes during the long, cold winter, as families in northern North America could not do without the heat for the hours it took to clean the pipes, and it was difficult to remove the creosote outside in freezing conditions.[37] If the creosote ignited inside the stove pipe, there would be "heart stopping panic in the house": the vents in the furnace would be closed, and the red hot pipes would be cooled with wet towels.[38] Sometimes the creosote was ignited in a planned burn designed to reduce the risk of an accidental fire. As one rural Canadian dweller described it, revealing in the process some of the gendered dynamics of energy anxiety: "If my mother would let him do it – or couldn't prevent him – [my father] would … purposely over-fuel the stove until the soot inside the stovepipes caught fire. Needless to say, she did not approve, but Dad laughed and said not-to-worry."[39] As another rural dweller remembered from her childhood, during such a planned conflagration, "My mother would have the baby in her arms and the young children about her, ready to run outside" in case the walls, often insulated with sawdust or newspaper, caught on fire.[40]

Explosions

The advent of fossil fuels in the home introduced women to the increased risk of explosions, with which these volatile combustibles quickly became associated. Women were afraid of exploding coal oil lamps as well as the (unwanted) fires they could cause. The author of the 1882 Toronto *Globe* article "Why Kerosene Lamps Explode" (as we saw earlier) began by reassuring his readers that kerosene itself was not flammable, and would never explode, unless adulterated with benzine or gasoline either through poor refining practices or when added by unscrupulous merchants. While harmless on its own, kerosene was, however, highly explosive when exposed to air. Unfortunately, as the author patiently explained, such exposure was quite likely to occur: "explosions generally occur when the lamp is first lighted without being filled, and also late in the evening when the fluid is nearly exhausted." When the light was extinguished at the end of the evening, "the lamp cools, and a portion of the vapour is condensed; this creates a partial vacuum in the space, which is instantly filled with air." As he goes on to explain,

> The mixture is now more or less explosive, and when, upon the next evening, the lamp is lighted without replenishing the oil, as is often done, an explosion is liable to take place. Late in the evening, when the oil is nearly consumed, and space above filled with vapour, the lamp cannot explode so long as it remains at rest on the table. But take it in hand, agitate it, and carry it into a cool room, the vapour is cooled, air pushes in, and the mixture becomes explosive.[41]

As the British Columbia Electric Co. noted in its self-promoting 1904 pamphlet, *A Chat on Electricity*, "the old way of lighting, by means of coal oil lamps is both dangerous and unsatisfactory: it is dangerous under any circumstance when being handled by either young or old, because a draught, or a slip of the hand, resulting in an upset, will instantly cause an explosion, with often terrible results."[42]

Many incidents of such kerosene explosions appeared in newspapers, informing women of the particular risks of the fuel.[43] The most commonly reported cause of kerosene explosions, however, was the unfortunate tendency for women to use kerosene as a fire starter in stoves. As Elizabeth Rogers explained, when she was a child in the 1940s, her mother's daily routine involved setting the fire in the wood stove each night. The next morning, "if it didn't go, she would put on little bit of kerosene if the wood was wet; even though it was dangerous."[44] No accidents resulted from this practice in the Rogers household, but others were not so fortunate, as numerous newspaper stories confirmed.[45] As the *Calgary Weekly Herald* noted with an apparent attempt at humour, "An exchange says that kerosene will remove rust from stoves. Unskilled persons have already discovered that kerosene will not only remove rust but incidentally the stove and occasionally the hired girl."[46]

Kerosene-fuelled stoves became popular in the early twentieth century, in part because they provided a small and focused fire for cooking in homes that were not connected to an electrical or gas grid.[47] They had the added advantage that they did not heat the entire kitchen. The insurance industry expressed concerns that they were a particular fire hazard because of their common use in summer kitchens, which were "generally board sheds with a sheet iron stove pipe (often rusted with holes) struck through the wood roof."[48] But as Carlotta Greer explained in her 1937 home economics textbook, accidents occurred most frequently by filling the kerosene stove when it was lighted: "No matter how hurried one may be, this should never be done. Always turn out the kerosene burners and let them cool before filling the stove. Of course, it is very inconvenient to do this in the midst of getting a meal."[49] According to American home economist Edith Allen, in her influential *Mechanical Devices in the Home*, housewives needed to follow instructions on their use exactly, for oil stoves were "not fool proof and should never be used by those who are afraid of them and who do not understand them."[50] The stoves needed to be kept clean and the wicks properly adjusted to prevent the fires associated with the build-up of soot, smoke, and gas. While the "oil stove cannot explode unless gas has formed in some part, like the tank, and becomes ignited by heat or spark," Allen continues, gas was, however, "more likely to collect in the tank when it is almost empty."[51] This statement was followed by a less reassuring section, "When the Stove Gives Trouble":

> In case the stove begins to blaze and cannot be controlled by the valves, remove the tank and carry it to some safe place where the kerosene cannot catch fire. When this is done, there is less than a pint of oil left in most stoves, and this will soon burn out without doing much harm, if clothing and water are kept away from the blaze. Open windows and doors to let out gases and smoke. If necessary, move the stove away from walls or furniture … oil stoves cannot explode when the tank is removed."[52]

The new fuels clearly had to be treated with caution, and it was essential for home safety that women understood their new and highly exacting requirements, and carefully monitored the appliance, in order to prevent potentially serious consequences.

But even the cast iron wood and coal stoves, themselves products of new fossil-fuel smelting processes, were not free from the risk of explosion.[53] As T.M. Clark patiently explained in his 1903 *Care of the House*, these time-tested stoves were in fact very unlikely to create a "really dangerous explosion" – unless of course someone lit a fire when the pipes between the stove and water tank were frozen or otherwise blocked: "many an unfortunate servant has been killed, and many a kitchen wrecked, by the terrific explosion which is sure to follow such carelessness in this respect."[54] When a coal stove exploded in a Pittsburgh home in 1870, setting the kitchen on fire, blasting the door off and breaking the cook's leg, the newspaper story noted that although it was naturally assumed that the explosion was due to the woman "cook's failure to turn on the water valve to the range's reservoir before she lit the fire," this was not the case in this incident; the most probable hypothesis was that "an obstruction of some kind in the pipes" that lead to the reservoir had caused a "dangerous build-up of pressure."[55]

But it was gas appliances that seem to have given women most concern about explosions. As critics from the newly emerging electrical industry never tired of pointing out, leaks from broken or improperly installed gas pipes and fixtures were very common, and the risk of explosion was always present when relying on gas appliances.[56] Furthermore, an improperly closed stopcock, left open after the flame was extinguished, like any other gas leak inside the house, could also cause a dangerous build-up of gas, creating the potential for deadly explosions. Household manuals sensibly cautioned that "a leakage of gas should never be investigated with the aid of a lamp, a candle or a lighted match. Innumerable explosions, many of them very destructive, have been caused this way."[57] Gas trade journals and sales catalogues, on the other hand, claimed that accidents from gas lighting were extremely unlikely unless the housewife were guilty of "mismanagement" by neglecting to ensure that burners were kept free of dirt and residue build-up, or by turning up the lights to too high a flame.[58] Whatever the causes, Canadian newspaper readers could look forward to regularly reading about any and all explosions relating to gasworks, gas stoves, or gas lighting across the continent and beyond.[59]

The gas stoves appearing on the market for the first time in the 1880s were a new cause for worry. Demonstrators had a difficult task explaining to potential customers how safe and reliable their products and services were, for "the knowledge that gas was explosive and toxic if inhaled instilled a prejudice that was difficult to dispel. This was particularly evident with reference to the gas cooker since it was generally held that food cooked in a gas stove was equally poisonous."[60] As Allen explained in *Mechanical Devices in the Home*, "many accidents happen when a stove is being lighted: The gas catches fire or explodes when the oven burner is lighted, blowing the oven door open or off the hinges, flashing out of the oven and burning any person near the stove." She prudently recommended opening the oven door several minutes before lighting the flame, to allow any concentrated gases to dissipate.[61] While gas companies maintained that as long as proper care and attention were exercised, their appliances were perfectly safe, evidence suggests that they were well aware of their dangers. Even after technological developments in the 1880s (particularly the gas mantle) made gas stoves safer, British doctors became alarmed by a perceived increase in stove-related domestic injuries and "warned householders of the dangers of improperly maintained gas stoves."[62] As Anne Clendinning has shown, accidents played a role in prompting gas companies to hire professional demonstrators, a "combination of sales assistants, domestic economy teachers and district health visitors."[63] The companies "anticipated that the use of educated female instructresses could minimize stove misuse, reduce the risk of household accidents and avoid future litigation and the resulting negative publicity."[64]

Interior Air Pollution

In addition to the risks of fire and explosion, a third source of considerable energy-related fear and anxiety for women was from what was generally called insalubrious, poisoned, or vitiated air. The bad smell of burning both manufactured gas and, later, kerosene (paraffin) raised serious concerns in an era in which smell was directly related to disease, but concerns about interior air quality extended far beyond toxic fumes associated with fossil fuel burning. Beliefs about pure air were related to the very intimate relationship that, historians argue and biologists support, exists between people and the environments that they inhabit. As Conevery Bolton Valencius explained it, in the nineteenth century people did not believe that "the environment stopped at the seeming boundary of the skin ... [T]he surrounding world seeped into ... every pore, creating states of health that were as much environmental as they were personal."[65] Melanie Kiechle and Nancy Tomes are among those who have documented the extensive discussions about fresh air and foul smells that were "simultaneously discussions of environment, health, material progress and the perils of urban life" throughout the nineteenth century.[66] Miasmic theories of disease – the belief that disease was caused by bad smells – gained increasing urgency as people struggled to understand the increased levels of contagion, disease, and epidemic illness in the rapidly growing and increasingly crowded cities.[67] Public health authorities were established in cities across North America and Europe, and gained new powers to regulate and contain "stench nuisances" on the streets, in industry, and in people's homes. By the 1860s, healthy air had become a lively cause for concern amongst the women responsible for keeping the home safe and healthy.

Medical journals, public health organizations, newspapers, household manuals, advertisers, women's organizations, and housewives themselves discussed the issue of air deeply and frequently: its freshness, its adequate circulation through appropriate ventilation, and the new dangers from the accumulation of sewer gases were all potentially sources of anxiety. As the chief plumbing inspector reported to the Third Congress of the Canadian Public Health Association in 1913, unhealthy air could come from a variety of sources, including "sewer gas, from the effects of trade [especially abattoirs] and suspended matter, such as road dust, seeds of plants, solid carbon, spores, germs and bacteria from organic matter."[68] As Frank Shutt pointed out in his 1893 article "The Air of Our Houses," vitiated air is "extremely deleterious to health … fainting fits, giddiness, nausea, and headache are recognized as the immediate results of breathing the air of badly ventilated halls and rooms, but it is not so widely known is that indigestion, diarrhoea and impaired and feeble condition of the system" were also common effects.[69] There were varying nineteenth-century explanations of exactly how air was involved with personal health, but by the 1860s, there was broad agreement about the healthful effects of well-circulating fresh air, and the deleterious effects of non-circulating fresh air.[70]

In Britain, coal burning fireplaces and stoves had long been associated with the creation of poisonous fumes. Indeed, it was the poisonous nature of coal smoke compared to wood smoke that led to the use of chimneys in London from the late Middle Ages, an architectural trend that spread across Britain as coal became a common form of heating and cooking from the seventeenth century onward.[71] By the early years of the nineteenth century, coal smoke, much of it from domestic heating and cooking, was polluting the air of cities, turning people's clothes and their faces black with the grime, and, as Petra Dolata argues in this volume, creating substantial work for the women tasked with keeping the home, and the people and objects in it, clean. As a result, its negative effect on respiratory health became the subject of frequent speculation.[72] Nevertheless, discussions of interior air pollution from carbon-based fuels before the 1860s focused overwhelmingly on the manufactured gas used for lighting, and later cooking. When gas lighting, and later kerosene, were first introduced into homes in the early nineteenth century, people were concerned about the bad smells and grime that accompanied the new forms of lighting. As Gooday has shown, many households were reluctant to install it for that reason, or they limited its use to hallways and kitchens.[73] The most extreme concern was about carbon monoxide poisoning, which could and did take lives on a regular basis, both at industrial sites and at home.[74]

While the gas companies "blamed the competing electricity industry for spreading false health rumours about the dangers of gas" and "consistently emphasized the cleanliness and comfort of gas," householders continued to complain that "coal gas was a dirty, smelly fuel which discoloured interiors, damaged textiles and required adequate ventilation to prevent the build-up of sulphurous fumes which caused headaches and nausea."[75] Despite assurances from the gas companies of the beneficial "influence of gas in promoting ventilation and circulation of the air and in sterilizing the air through the cremation of bacteria in the bunsen flame,"[76] complaints about the "vitiating effects" of gas persisted.[77] The Regina Plumbing Inspector reported that he had been called to look for a possible sewer gas leak in a house where the family blamed their "constant sore throats" on the sewers, but their ailment ceased once "a hood and a flue had been fixed over the gas stove."[78] There were the "hundreds of deaths each year," according to T.M. Clark in *The Care of the Home*, that were caused by air poisoned by undetected gas leaks, with the result that people "perish later in their sleep through the slow diffusion of the poison in the room."[79]

But by the 1860s, coal fires, gas stoves, and gaslight became associated with a more specific risk: carbonic acid. It is a difficult concept to define, as it does not actually exist. Combining the qualities of carbon dioxide in terms of human respiration, and carbon monoxide in terms of carbon combustion, carbonic acid became a focal point of serious concern among public health officials, doctors, and women. From the vantage point of the twenty-first century, where so much more is now known about the harmful effects of fossil fuel burning on human health, the concept of carbonic acid represents a curious interface between people's concerns about their bodily health, air pollution, and their consumption of carbon-based fuels.

Carbonic acid was not originally identified with coal- or gas-burning appliances, though discussion of it coincided with their increasing use in nineteenth-century homes. It was originally identified as the organic polluting agent in vitiated air, and the scourge of indoor environments in crowded and under-ventilated places. Carbonic acid gas was regarded as "the chief impurity ... of vitiated air, and the one constituent that it is necessary to determine when examining an air for hygienic purposes."[80] If crowded public spaces could be dangerous, so too were private bedrooms, where all too often there was not sufficient circulation of air to cleanse it of the impurities that accumulated overnight.[81] Theories about what exactly carbonic acid was seem to have drawn on new scientific studies of human respiration that demonstrated that carbon dioxide was a waste product of breathing, though carbonic acid was distinguished from carbon dioxide. Shutt explained that carbonic acid was "the result of the union of carbon (or charcoal) with oxygen. It is formed by the process of combustion, in the respiration of animals, and by decay or putrefaction of organic matter in the air." While sanitarians had previously distinguished the harmful miasmic vapours emanating from decaying organic matter from chemical pollutants such as coal smoke, Shutt explained that "The chemistry, as far as the result is concerned, is precisely the same in all of these. The burning of wood, coal or other material rich in carbon and hydrogen is accompanied by the development of heat and light. That is what is commonly understood as combustion." The heat generated by the combustion of food in the body, Shutt further explained, converted "to carbonic acid and aqueous vapor, is precisely equal to the amount that would have been produced if the food material had been burnt in the air."[82] Carbonic acid, then, was the result of "animal exhalations," and explains the "heavy, sickening smell noticed on first entering a crowded room ... Of all the noxious matters in the fouled air of a poorly ventilated school or public building, they are at once the most perceptible, the most offensive, and the most rapidly prostrating."[83]

In the face of the growing realization that "the breath of man is deadly for his fellow creatures,"[84] scientists sought to quantify the exact amount of carbonic acid produced. The Canadian *Sanitary Journal* reported in 1874 that "children of average school age throw out each, by respiration, about three gallons per hour of poisonous gas, animal impurities, and watery vapor; and that in every 1000 gallons of these deleterious substances are 3 gallons of dead, decomposing animal matter."[85] The same journal reported that:

One cubic foot of ordinary atmospheric air of average purity, contains less than one cubic inch of carbonic acid. One cubic foot of expired air contains over 70 cubic inches of carbonic acid. The average amount of carbonic acid exhaled from the lungs of an adult, under ordinary circumstances, in 24 hours, is about 16 cubic feet; in which are about 7 1/2 ounces of solid carbon.[86]

Though there seemed to be some confusion about whether the deleterious effects of vitiated air was due as well to a relative absence of healthful oxygen or to other polluting content in the air, discussions of the dangers of excessive carbonic acid made frequent reference to the contributions that coal-burning stoves and gas lighting made to the carbonic acid content of people's homes. As the *Sanitary Journal* cited above went on to note, "the combustion of 1 cubic foot of coal gas gives rise to 2 cubic feet of carbonic acid; while it consumes the oxygen of 10 cubic feet of air. The combustion of 1 pound of oil produces about 21 cubic feet of carbonic acid."[87] As William Baker phrased it in his 1860 book on ventilation, the home was indeed a particular topic of concern:

> nothing connected with that home, its health, comfort and happiness, can justly command the important considerations that are connected with the purity of the air – the element upon which chiefly depends the existence of those most dear to us, our wives and children, who spend the greater part of their lives within doors. Hence the artificial heat, which may make that air impure, becomes a subject of paramount significance.[88]

While miasmic theories of disease help to explain why fresh, circulating air was identified as healthy, Baker's last sentence here reveals the connection being made between heated air and the perils of carbonic acid. Catharine Beecher and her sister Harriet Beecher Stowe comment authoritatively on the subject in their widely read *The American Woman's Home*: "The subject of the ventilation of our dwelling-houses is one of the most important questions of our times."[89] They quote a Doctor John Griscom, author of the *The Uses and Abuses of Air*, who in his examination before public health commissioners in Great Britain declared: "Deficient ventilation I believe to be *more fatal that all other causes put together*." The authors go on to ask, "How many thousands are victims to a slow suicide and murder, the chief instrument of which is want of ventilation! ... No wonder there is so much impaired, nervous and muscular energy, so much scrofula, tubercles, catarrhs, dyspepsia and typhoid disease."[90] Doctors, sanitarians, health inspectors, and women became increasingly concerned that the carbon-based fuels they were burning in their homes to provide light and heat were also poisoning the air, by means of the carbonic acid that they created. As the authors of *The American Woman's Home* explained, "carbonic acid is formed by union of oxygen with carbon or charcoal ... [I]f received into the lungs undiluted by sufficient air it is a fatal poison, causing certain death. When it is mixed with only a small portion of air, it is a slow poison, which imperceptibly undermines the constitution."[91] In its 1875 article "Carbonic Acid and Its Fatal Effects," the *Sanitary Journal* provided a worrying but familiar catalogue of cases of asphyxiations, declining health, and acute health issues, including cholera, caused by carbonic acid.[92] As Beecher explained in *The American Woman's Home*:

> Tight sleeping rooms, and close, air-tight stoves are now starving and poisoning more than one half of this nation … It seems impossible to make people know their danger. And the remedy for this is the light of knowledge and intelligence which it is woman's special mission to bestow, as she controls and regulates the ministries of the home.[93]

In 1868, the *Canadian Pharmaceutical Journal* quoted an "Expert Physician" in France who "does not hesitate to assert most positively that cast-iron stoves are sources of danger to those who habitually employ them."[94] Unlike a roaring fire in the open hearth, "stoves did not encourage the healthful flow of air through the house and up the chimney." Furthermore, iron and cast iron, "when heated to a certain degree, become pervious to the passage of gas … the air which surrounds a stove is saturated with hydrogen and oxide of carbon" giving rise to carbonic acid.[95] Compounding the problem, "stoves with defective flue pipes, stoves and lamps which have no flue connections deprive us of a considerable proportion of oxygen of the air."[96]

Conclusions

Women's worries and fears for their family's health and safety were exacerbated by the novelty, the unpredictability, the strange odours, and the extreme flammability of the new fossil fuels in the home. In the absence of reliable information about new forms of energy, such concerns comprised an important and complex part of their energy-related decisions, decisions that would eventually forge new relationships between women and the environment, and between women and the wider society. Energy fear and anxiety have not received much attention from historians, nor has the gendered nature of energy transitions. The tendency of historians to tacitly ignore women's energy fears in the past exists in stark contrast to the rapidly growing medical and scientific evidence documenting the devastating impact that fossil fuels have demonstrably been having on both the environment and on human health since this period, and in ways that continue to place particular burdens on women.[97] From the acidification of the oceans and climate change, to oil spills and the toxic, polluted air that is now implicated in at least seven million premature deaths each year, fossil fuels are taking a terrible toll on the planet.[98] While no one now believes in the perils of carbonic acid, it turns out that there have always been grave reasons to fear the consequences of using fossil fuels in and out of the home. Fires, explosions, and air pollution from these energy sources have had a devastating impact on the planet and almost everything living on it.

Can we make anything at all from the fact that people were right for the wrong reasons? Probably not. From the vantage point of the twenty-first century, it is clear that women's fears and anxieties about fossil fuels, though not ungrounded, were misdirected. Right for the wrong reasons, women did little, however, to politicize their specific energy-related fears and anxieties in this period.[99] On the contrary, rather than organizing around their fears to advocate a limit to fossil fuel use, or to put the onus on corporations and governments to improve home safety, as they later would do, the home economics movement in this era instead organized women so as to make them more scientifically and technically knowledgeable, and more rational in their energy use.

But the particular response to fossil fuels that this paper has documented – to fear fossil fuels but to continue, for the most part, to use and negotiate with them – seems particularly relevant as we contemplate our own responses to energy-related danger in the present. Citizens around the world are mobilizing around their fears of climate change to argue the necessity of another energy transition, to a post–fossil fuel world. The direct links between fossil fuels and human health, particularly that of children, is also being discussed with increasing urgency: as the director general of the World Health Organization recently put it, air pollution is the "new tobacco." The simple act of breathing, he wrote, is killing seven million people a year and harming billions more, but "a smog of complacency pervades the planet." Unconsciously echoing the nineteenth-century crusaders for pure air, he concluded, "A clean and healthy environment is the single most important precondition for ensuring good health."[100]

Historical studies like this one (and indeed like all of those in this volume), that recognize the ambiguities and contradictions in people's attitudes and behaviours, that document differences between what women said and what they did, and that try to find the social and cultural contexts for understanding and explaining these dissonances, may provide some insights into the ways in which women remain deeply connected to the environment through their energy use. Perhaps these can help to frame some questions that might be important in negotiating future energy transitions, such as, how do we account for the levels of danger that people, including women, are prepared to live with? What are the cultural, political, and economic factors that have stoked and assuaged those fears? And perhaps most importantly, what do women and men need to know, and believe, in order to use fear and anxiety to promote change?

Notes

1 See, for example, Parr, *Domestic Goods*; Schwartz Cowan, *More Work for Mother* and "The Consumption Junction"; Gooday, *Domesticating Electricity*; and Goldstein, *Creating Consumers*.

2 For the slow uptake of modern appliances in this era in Canada, see Sandwell, "Pedagogies of the Unimpressed," and in Britain, Gooday, *Domesticating Electricity*.

3 Rachel Plotnick and others have argued that, "Due to the anxiety that this new form of power [electricity] often produced, many sectors of society intervened to provide education that would make electrical activities more intelligible and acceptable." Plotnick, "At the Interface," 838. See as well, Clendinning, "Deft Fingers and Persuasive Eloquence," and Sandwell, "Pedagogies of the Unimpressed."

4 Tarr, "Housewives as Home Safety Managers," 200.

5 The term "interface" is drawn from Rachel Plotnick's thoughtful discussion of people's first experiences of electrical energy use. Plotnick, "At the Interface." See also Harrison Moore, this volume.

6 Not discussed here is another major source of anxiety for mothers: the dangers that automobiles posed to their children in the 1910 to 1939 period – a concern that was well-deserved. This under-researched topic deserves a chapter of its own. For a discussion of the "slaughter" of children on American city streets by trolley cars, trams, and, later, cars and trucks, see Zelizer, *Pricing the Priceless Child*.

7 For other interpretations of energy danger and anxiety, see Zallen, *American Lucifers*; Simon, *Dark Light*; and Hunt, "Anxiety and Social Explanation." Environmental historians are documenting the considerable toll that fossil fuel consumption has taken, directly and indirectly, on human health over time. See, for example, Floud, et al., eds., *The Changing Body*.

8 This term to describe the home was popularized by Christopher Lasch's bestselling book of that name. *Haven in a Heartless World: The Family Besieged* (1977) was one of the first cultural critiques of the modern family, criticizing the ways in which modern experts were replacing foundational moral and social roles of the family.

9   Boydston, *Home and Work*; Parr, "What Makes Washday Less Blue?";
    Schwartz Cowan, "The Consumption Junction" and *More Work for Mother*;
    Sandwell, "Pedagogies of the Unimpressed"; Davidoff and Hall, *Family
    Fortunes*; Strasser, *Never Done*.

10 See, for example, Melosi, *The Sanitary City*; Tomes, *The Gospel of Germs*;
    and Kiechle, *Smell Detectives*. For the British case, see, for example, Luckin,
    "Revisiting the Idea of Degeneration in Urban Britain, 1830–1900."

11 Tomes, "The Private Side of Public Health." On "house disease" in Britain,
    see Adams, *Architecture in the Family Way*.

12 Tarr, "Housewives as Home Safety Managers," 200.

13 Tomes, "The Private Side of Public Health," 510.

14 These are nicely summarized in Tomes, *The Gospel of Germs*; Kiechle, *Smell
    Detectives*; and Strasser, *Never Done*.

15 Goldstein, *Creating Consumers*. The essays in Stage and Vincenti, eds.,
    Rethinking Home Economics provide a good overview of the range of women's
    activities, beliefs, and aspirations included in that broadly based movement.

16 These include the role of mass consumption in boosting energy
    consumption, and the exigencies of expensive centralized network systems
    of energy delivery, where costs come down the more people "buy in" to the
    network. Another reason women's energy use in the home was a cause of
    national concern related to fears about rural depopulation, and on the other
    hand that modernizing improvements in society more generally would be
    rejected if women did not adopt "modern living" in their homes. These are
    explored in detail in the Canadian context in Sandwell, "Pedagogies of the
    Unimpressed," and are nicely summarized in the American context in Tobey,
    Technology as Freedom.

17 For an overview of the triumphalist narrative in the British case, see for
    example, Luckin, *Questions of Power*; for the Canadian, Parr, "Modern
    Kitchen, Good Home, Strong Nation"; and Sandwell, "Pedagogies of the
    Unimpressed"; and the American, Kline, "Agents of Modernity" and Jellison,
    Entitled to Power.

18 Gooday, *Domesticating Electricity*, 5, 6; Sandwell, "Pedagogies of
    the Unimpressed."

19 Tarr, "Housewives as Home Safety Managers."

20 "Why Kerosene Lamps Explode," *Globe* [Toronto], 4 September 1882, 3.

21 "How to Overcome Fear of Electric Wringer," *Globe*, 3 September 1920, 12.

22 Comment, *London Advertiser*, 20 July 1880, reprinted from the *New York
    Tribune*, nd.

23 Beynon, "Women's Clubs," 25. As Joy Parr points out, "Women working at
    home for free were sensitive to criticisms that household tasks required
    little skill. They were not won over by claims about 'how little they need do
    themselves' once they had a new range, partly as a woman marketer noted,

'because they resent being told that a product can do their job better and quicker than they can.' An appliance which 'saved' the labour of a woman who did not work for wages cost, but did not save, money.' Parr, *Domestic Goods*, 205–6.

24 See as well, Sandwell, "The Coal-Oil Lamp."

25 Sandwell, *Heat, Light and Work*, interview no. 3.8.

26 Hammond, *The Electric Light in Our Homes*, 59–60.

27 For the dangers of exploding camphene, see Williamson and Daum, *The American Petroleum Industry*, vol. 1, 47–8; and O'Dea, *The Social History of Lighting*, 56. For the best history to date of camphene lighting, see Zallen, *American Lucifers*, chapter 2, "Piney Lights," 57–93.

28 For an overview of the dangers of kerosene lamps, see Wlasiuk, *Refining Nature*, 44–8. See also, for example, *London Advertiser* [Ontario], 28 May 1880; *London Advertiser*, 20 July 1880; and "The Kerosene Lamp: The Panic and Disaster that Followed the Upsetting," *Globe*, 24 November 1885.

29 In 1941, almost 80 per cent of Canadian farm homes were relying on kerosene lighting, and and 31 per cent of all homes. Across the vast spaces of the rural western Canadian prairies and in Prince Edward Island on the east coast, a majority of all homes were still relying on coal oil lamps by that date. Sandwell, "The Coal-Oil Lamp," 196–7.

30 Wlasiuk, *Refining Nature*, 46–7.

31 Thomson and Redwood, *The Petroleum Lamp*, 52–4.

32 "Fires on the Farm," in *Monetary Times, Trade Review and Insurance Chronicle*, 1855.

33 *Scientific American* 42, no. 7 (1880): 102.

34 Tarr, "Housewives as Home Safety Managers," 208.

35 Ibid., 209. As Tarr explains, the National Safety Council was founded in the US in 1914, but it focused almost exclusively on industrial safety and automobile accidents until 1940, when home safety became a topic of increasing concern.

36 Harris, "Conquering Winter," 41; Harris, "Inventing the US Stove Industry"; MacDonald, "How the Cooking Stove Transformed the Kitchen," 5; Brewer, *From Fireplace to Cookstove*, 170.

37 "Fires on the Farm" and "Our Fire Appliances" in *Monetary Times, Trade Review and Insurance Chronicle* 28, no. 42 (19 April 1895): 1855.

38 Sandwell, *Heat, Light and Work* interview, no. 4.7. See also Kathleen Tobin interview, "Behind the Kitchen Door Project," File 4008:0014, British Columbia Archives.

39 Sandwell, *Heat, Light and Work* interview, no. 1.8

40 Ibid., interview, no. 4.5

41 "Why Kerosene Lamps Explode," *Globe*, 4 September 1882, 3.

42 British Columbia Electric Co., *A Chat on Electricity*, [c. 1904], Royal British Columbia Archives and Library, NWP 971.91, B863.

43 See for example, *London Advertiser*, 28 May 1880; "Waterloo, Burned to a Crisp," Toronto *Globe*, 7 December 1880, 8; 10 February 1882.

44 Interview with Elizabeth Rogers, "Behind the Kitchen Door Project," file T4088:006, 007, British Columbia Archives.

45 *London Advertiser*, 28 May 1880; *Edmonton Bulletin*, July 22 1910, 2; *Edmonton Bulletin*, 30 March 1912, 1; *Edmonton Bulletin*, 17 July 1913, 8; *Edmonton Bulletin*, 16 August 1916; *Edmonton Bulletin*, 23 October 1908; *Wetaskiwin Times*, 15 December 1921, 1. See also, "As Usual," *Edmonton Bulletin*, 30 May 1907, 1.

46 *Calgary Weekly Herald*, 25 May 1899, 4.

47 *The Lamp*, September 1918, 24.

48 "Fires on the Farm" and "Our Fire Appliances" in the *Monetary Times, Trade Review and Insurance Chronicle* 28, no. 42 (19 April 1895), 1855.

49 Greer, *Foods and Homemaking*, 180.

50 E. Allen, *Mechanical Devices in the Home*, 31.

51 Ibid., 32–5

52 Ibid., 36.

53 J.H. Harris, "Inventing the US Stove Industry."

54 T.M. Clark, *The Care of a House*, 46–7.

55 *New York Times*, 27 December 1870, 2, cited in Brewer, *From Fireplace to Cookstove*, 170.

56 T.M. Clark, *The Care of a House*, 190–1.

57 Ibid., 190. Clark points out that it was quite easy to turn the stopcocks on the lighting fixtures so far as to reopen them after the flame was extinguished. In the darkness, it could be impossible to recognize the error (193). Lighting a match or other lighted flame to explore a gas leak was nevertheless a surprisingly common cause of gas explosions. See, for example, the following stories in the Toronto *Globe* over a five-year period: 1 September 1883, 4 (Toronto); 4 January 1884 (Montreal); 19 August 1884, 1 (Prescott); 6 December 1885, 2 (Montreal); 6 June 1887 (Toronto); 16 December 1887 (Winnipeg); 30 April 1888 (Owen Sound).

58 E. Allen, *Mechanical Devices in the Home*, 26.

59 See, for example, *World* [Toronto], 22 January 1884; 2 December 1884; 20 May 1885; 2 January 1886; 3 April 1886; 12 November 1886, and 3 January 1918.

60 Clendinning, "Deft Fingers and Persuasive Eloquence," 504.

61 E. Allen, *Mechanical Devices in the Home*, 26.

62 Clendinning, "Deft Fingers and Persuasive Eloquence," 509.

63 Ibid., 502.

64 Ibid., 509. These worries are confirmed in the oral history record: see Sandwell, *Heat, Light and Work*, interview no. 2.9; and Mary Butler, Interview, "Behind the Kitchen Door Project," British Columbia Archives, T48088:0021.

65 Valencius, *The Health of the Country*, 12. For environmental historians' confirmation of these relationships, see Floud, et al., eds., *The Changing Body*.

66 Kiechle, *Smell Detectives*, 15.

67 Kiechle, "Navigating by Nose," 764. On early responses to oil and gas pollution, see Rosen, "Knowing Industrial Pollution"; and Mosley, *Chimney of the World*.

68 Mathias, "Leaves from an Inspector's Note Book," 563.

69 Shutt, "The Air of Our Houses," *Ottawa Naturalist* 7, no. 2 (May 1893): 30.

70 For a discussion of the ways in which this intersected with architecture, see A. Adams, *Architecture in the Family Way*; and with air conditioning, see G. Cooper, *Air-Conditioning America*.

71 As Robert Allen argues, "switching fuels … presented complex design problems." Domestic fires were originally in the centre of the room, away from the walls, and smoke exited through a hole in the roof. But because coal smoke is poisonous, and needs a small enclosed space for efficient combustion, it requires specialized equipment – including a chimney to remove the fumes, which led to substantial architectural reforms, particularly after the fire of London in 1666. Allen, "The Shift to Coal," 13–14.

72 For an excellent discussion of the implications of coal burning for everyday life, including for women, see Mosley, *The Chimney of the World*, although Mosley's key question is why, given the huge negative effects that living with smoke had on the people in and around Manchester in the nineteenth century, were there not more attempts made towards its abatement?

73 Gooday, *Domesticating Electricity*.

74 Many of industrial accidents were related to the gas industry. See, for example, *Intercolonial Gas Journal of Canada* 16, no. 3 (March 1923), which reported ten men dying from carbon monoxide poisoning. See also Sandwell, *Heat, Light and Work in Home*, interview no. 1.8.

75 Clendinning, "Deft Fingers and Persuasive Eloquence," 504.

76 "Chimneys for Gas Stoves," *World*, 9 June 1884; "The Sale of Gas for Illumination by R.F. Pierce," *Gas Industry: Heat, Light, Power* 13, no. 1 (1913): 64–5. The gas companies blamed the competing electricity industry for spreading false health rumours about the dangers of gas.

77 Hammond, *The Electric Light in Our Homes*, 45.

78 Mathias, "Leaves from an Inspector's Note Book," 563.

79 T.M. Clark, *The Care of the House*, 194.

80 Shutt, "The Air of Our Houses," 27–8.

81 Ibid., 29–30; "Bed-room Ventilation," *Sanitary Journal* 1, no. 3 (November 1874): 87–8.

82 Shutt, "The Air of Our Houses," 27–8.

83 "School Room Ventilation," *School Magazine*, September–October 1881, 232.

84 "School House Ventilation," *Sanitary Journal* 1, no. 2 (September 1874): 57.

85 Ibid.

86 "Facts and Figures," *Sanitary Journal* 1, no. 2 (September 1874): 60–1.

87 Ibid.
88 Baker, *Artificial Warmth and Ventilation*.
89 Beecher and Beecher Stowe, *The American Woman's Home*, 62.
90 Ibid., 53, 62. Emphasis in the original.
91 Ibid., 47, 35.
92 "There can be no doubt that death in cholera is chiefly due to poisoning by carbonic acid." "Carbonic Acid and Its Fatal Effects," by W.H. Thayer, *Sanitary Journal* 1, no. 4 (January 1875): 101–3.
93 Beecher and Beecher Stowe, *The American Woman's Home*, 36.
94 He presented evidence from a recent epidemic in Savoy, "where everyone who used an imported iron stove died," while those using other "modes of firing or other sorts of stoves" were left "untouched by the disease." *Canadian Pharmaceutical Journal* 1, no. 7 (November 1868): 106.
95 Ibid.
96 Mathias, "Leaves from an Inspector's Note Book," 563.
97 Fraser, "Behind Marx's Hidden Abode."
98 "Air Pollution Is the 'New Tobacco,' Warns WHO Head," *Guardian*, 27 October 2018.
99 Mosley's *The Chimney of the World* provides an excellent cultural history of why there was so little political opposition to intense coal smoke pollution in nineteenth-century Manchester. As other authors in this collection argue, this was not universally the case.
100 "Air Pollution Is the 'New Tobacco,' Warns WHO Head," *Guardian*, 27 October 2018.

4

Agency, Ambivalence, and the Women's Guide to Powering Up the Home in England, 1870–1895

Abigail Harrison Moore

Introduction

The Industrial Revolution drove middle-class women into a newly imagined private sphere – the home. In England, as the country's wealth increased exponentially, those who benefitted from this new money looked to invest in their homes, often following the example previously laid down by the country house and town house owning aristocrats, whose power and influence did not wane with the growth of the new "middle" class. These homes became the province of the middle-class "housewife," who was expected to manage all aspects of these increasingly elaborate spaces, crafting a safe, comfortable, tasteful, and moral interior from which to welcome her husband back on his return from work and to stage the family's social standing.[1]

Industrialization has been seen by many historians of this period as enhancing gender inequality, particularly for these middle-class wives and mothers. They have argued that socially created expectations of gender roles saw wives "limited" to the home in a society increasingly dominated by the separate spheres for men and women of work and home, but in this chapter I want to explore how the rapidly growing expectations of women as consumers in the home led them to be very active in energy decisions just at the moment when the transition to domestic electricity or gas became a possibility in the 1870s and 1880s. The women I will focus on, the newly emerging decorators and advisors, found an opportunity for paid and professional work as a direct result of the "ravenous ... appetite for artistic instruction" amongst the housewives tasked with producing a beautiful home.[2] Importantly, these women combined a call for women to have agency in domestic decision-making with a wider call for suffrage. In this chapter I aim to link their pioneering advice on lighting the home with their pioneering roles as women finding their voices in the patriarchal milieu of Victorian England.

As women were increasingly expected to provide the perfect home, and perform their duties as "the angel in the house," they needed to turn to others for guidance on how to navigate the rapidly expanding market of choice for interior decoration and design.[3] Into this space came one of the first professional middle-class women – the decorator – who framed herself, through selling advice and furnishings, and writing guides to interior design, as indispensable to the middle-class housewife wondering which way to turn when it came to lighting her home in the best possible taste and for the best possible effect. Women's first professional engagement in energy transitions was therefore often via art, craft, design, and decoration, a fact to date very much overlooked in the histories of energy. In this chapter, therefore, I also want to demonstrate what the history of art can bring to the history of energy, and in doing so, start to unpick the gendered reasons for why my discipline rarely features in the histories of gas and electricity, despite offering us an important way of understanding, through the design of lighting in the home, the social history of energy decision-making. One key barrier to this seems to be the need to look inside the middle-class home, an increasingly feminized world of beauty, aesthetics, and comfort, to find many valuable archives of the social histories of energy. Whereas historians have often dismissed the domestic world, Bonnie Smith is among the historians who have emphasized that "domesticity, and not captial, was central to nineteenth century life ... The writing of history will have to take the home as its focal point, not capitalization, industrialization, or politics."[4] In my exploration of the women who guided housewives on the aesthetics of their energy decisions in the home, my aim is to not simply reinsert middle-class women into energy history, but rather to empower them in this history by exploring their experiences and agency as professional decorators and the women they assisted, to offer an initial analysis of their suggestions when it came to lighting up the middle-class English home at the end of the nineteenth century.

This chapter will consider two case studies of the emerging profession of the decorator and their guidance on lighting of the home, with a precise focus on the first women who identified professionally as advice-givers.[5] Energy supply, like home design, had traditionally been considered a decision for the "master" of the house, and the historiography of energy histories has tended to follow this path. In considering the pioneering case studies of two advice guides written by women, for women – Agnes and Rhoda Garrett's *Suggestions for House Decoration* (1877) and Mrs Mary Eliza Haweis's *The Art of Decoration* (1881)[6] – this chapter will explore the role of the newly emerging professional women decorators and the ambivalence at the heart of their advice to middle-class housewives lighting up their homes. The Garretts, whom I will consider first, achieved a number of "firsts" for women: first to be apprenticed in an architect's office, first to be professional decorators, and first to publish a guide to decorating the home which included recommendations on gas lighting. While Mrs Haweis did not work as a professional decorator (although she was often invited to a "friend's" home to "offer advice"), she mirrored the Garretts by linking a successful career writing books on art, fashion, and the home to a wider call for women's enfranchisement. She was also a pioneer specifically in the history of energy, as she was the first woman I have found to date to recommend in print the use of electricity to light the English home.

Women Guiding Women in Energy Decisions

In 1877, when cousins Agnes and Rhoda Garrett published *Suggestions*, they spoke directly to the "ladies of the family" who "have their tastes consulted" when the "furniture and decorations ... of the house have to be chosen."[7] Clive Edwards has proposed that the Garretts, creators of the first all-female design and decorating company in Britain, used their experience to attempt to subvert the patriarchy of the household, evidenced by influential books such as Robert Kerr's *The Gentleman's House* (1864).[8] Kerr had successfully suggested to the country house owner that there should be clearly delineated gender divisions both in terms of the actual planning of the home and in the domestic decision-making. The Garretts, in contrast, targeted their publication directly at women readers and aimed to guide them as to the appropriate choices for the aspirant middle-class housewife.

Such guidebooks suggested that household décor was a public reflection of personal taste, and, as such, the fear of "getting it wrong" opened up the potential for professional decorators to "assist" them. The Garretts established that women responded well to being guided by women. In their attempts to move away from the concept of the amateur home decorator, they aligned themselves with the other professions.[9] They wrote that "decorators may be compared to doctors. It is useless to put yourself under their direction unless you mean to carry out their regime."[10] This quote demonstrates both the ambitions of the

Garretts and the ambivalence that lies at the heart of my analysis. This ambivalence is rooted in the fact that the women in my chosen case studies of advice-givers – the Garretts and Mrs Haweis – had two different and opposing messages. On the one hand, as professional decorators and guides they were preying on the insecurities, fear, and ignorance of women consumers in making their own decorating decisions in the 1870s and 1880s, when the transition to gas and electricity became a possibility in the home, by giving them "expert" advice. On the other hand, they were also explicitly claiming that such decision-making empowered their women clients, through linking women's energy-decorating decision-making to their need to free themselves from the subjugation of men. While their guidebooks exhorted women to see home decoration as an enfranchising activity, aligning this work with the politics of suffrage, they also capitalized on a woman's need for guidance as a way of opening up a client base and building their business. As such, the first women decorators walked a fine line between a deficit model of advice, embodied by the salesman's age-old pitch of "you need me because you cannot do this yourself," and positively encouraging women to find their own style as part of a wider opportunity to break away from patriarchal traditions.[11] Rather than framing these professional guides as being either about controlling the housewife or empowering her, I want to use my case studies of lighting advice to suggest a new way of understanding how women worked with each other through times of energy transition and change.

The Garretts' Suggestions for Women

Agnes Garrett was born in 1835 to a large and wealthy middle-class family that unusually for the time encouraged the education of women.[12] Two of her sisters became campaigners for women's rights: Millicent, who went on to lead the National Union of Women's Suffrage Societies, and Elizabeth, who was the first woman in Britain to qualify as a doctor.[13] In the 1870s, Agnes and Rhoda did several speaking tours together and were committed members of the women's suffrage cause. Rhoda argued to an audience at the National Association for the Promotion of Social Science in 1876 that,

> The woman's sphere and woman's mission is one of the most important problems of the present day, but here, at least, in the decoration and beautifying of the house, no one will dispute their right to work. If a woman would rightly undertake this work and

would study to understand the principles upon which ... it is based, they would not only thereby increase their own happiness, but in thus extending the gracious influence of the home, they would help raise the position of household art, and thus render a real service to the nation.[14]

Notwithstanding the trivialization of home decorating within male political discourses then and now, feminist historians have argued that women's growing agency in managing the nineteenth-century middle-class home not only must be seen within the larger contexts of reforms for women, it could also be seen as functioning as an agent of reform in itself.[15] The work of the Garretts certainly provides evidence that they believed that female empowerment was an important component of what they were selling with their services. The Garretts were starting out in their profession as decorators at a point when women were only just taking over as the main purchasers for the home. Hence, those wishing to sell different forms of energy use into their homes needed to start addressing them directly, including the authors of advice guides, such as the Garretts, Mrs Haweis, Mrs [Lucy Faulkner] Orrinsmith (*The Drawing-Room: Its Decorations and Furniture*, 1878), Mrs J.E. Panton (*Suburban Residences and How to Circumvent Them*, 1896), and Mrs J.E.H. [Alice] Gordon (*Decorative Electricity*, 1891), all of which I am considering as part of a larger project to understand the role of women and books on decorating the home in energy transitions.[16] The Garretts and Mrs Haweis, on whom I here concentrate, were pioneers as women decorators finding their way in what had traditionally been viewed as a man's world.[17] Of course, the period that I am looking at is right at the beginning of the story of electricity in the English home, but this is very much my point, as the women I am considering can be seen as trailblazers, not only in terms of fighting for professional roles for women, but also, in Mrs Haweis's case, in the history of domestic electrification.

While middle-class women had multiple duties and responsibilities in the home, this was rarely recognized due to their non-renumeration, and the question of middle-class paid work for women only started to be debated in the 1880s. For example, in 1888, an article on the Women's Arts and Industries section of the Glasgow International Exhibition commented that, "One of the pressing questions of the day is, how to turn to good account the energies and capabilities of the vast body of women who have no work to do."[18] Rhoda Garrett was desperate to become an architect, but was unable to find an architectural practice willing to take a "lady pupil."[19] The challenges and necessity of finding an apprenticeship in this profession that had heretofore been seen as being only open to men runs as a motif through the Garretts' approach to reforming women's work opportunities. Similarly, the Women's Arts and Industries section of the Glasgow International Exhibition aimed "to extend the knowledge of what women can do ... The competition in the world's business grows fiercer and fiercer every day, and only the men and women who are well equipped for the fight can hope to hold their own."[20] Hence the ambition to see art, design, and decoration as a suitable focus for middle-class women became part of the wider fight for suffrage, framed as of benefit for the British economy more generally.

Women's increasing power as consumers in the home reflected and influenced the wider re-evaluation of women's power and rights as part of the campaign for married women's rights, culminating in the Married Women's Property Acts in 1870 and 1882. Before the 1860s the choice of domestic furnishings was primarily a male activity.[21] Hence the furniture and interiors trade, often framed by the idea of the upholsterer who would provide everything for the home, being dominated by men up to this point. Whereas, as the case study of the Garrett cousins demonstrates, by the late 1870s, women were starting to provide guidance in this area, given that women were the intended customers of their guidance. For example, Moncure Conway stated that, "When decorative work of such firms as Messrs Morris and Co. began to attract general attention, it appeared that it offered opportunities for employment suitable to women."[22] That said, this was not an easy path to take. Conway went on to comment on the "struggle" that the Garrett women had to go through to obtain their positions, and the Garretts used this struggle to inform and feed their rhetoric and actions on women's rights.[23] In 1875, Agnes and Rhoda set up their business, "A & R Garrett House Decorators," from their home at 2 Gower St, Bloomsbury, London. Conway confirmed in 1882, shortly before Rhoda's death, that they were "an independent firm, with extensive business," which had "gained fame ... by their successful decoration of many private houses."[24] From early in their career they referred to themselves as "architectural decorators," an altogether more professional, more masculine label than "house decorator" that effectively dealt with their exclusion from the professional designation of "architect" by hinting at it.[25]

The aim of Suggestions, a "small treatise ... chiefly to do with the internal fittings and decorations of houses," was to provide "a few practical hints to those who are desirous ... to make their houses more pleasant and restful homes to live in."[26] Suggestions was the second volume in a series of "Art at Home" books (1876–1883), all written by "Lady Experts," described as "the professional advisers of the middle-classes,"[27] and one of the most successful, with six editions published by 1879 and 7,500 copies printed.[28] It was planned as "an account of the more simple ways in which, without great expense a home might be made pretty and also wholesome."[29] While the series as a whole was intended to address a certain class of people, those of "moderate or small income,"[30] the Garretts' volume spoke specifically to "the decoration and furnishing of middle-class houses,"[31] although many of their customers, including the Beale family, whom I have considered elsewhere, were certainly in a much higher wealth bracket.[32] These were precisely the sorts of customers who could afford to light and heat their homes with the new technologies of gas and electricity, but their wealth also gave them the freedom to choose and they needed to be persuaded to invest in both a certain type of energy and the aesthetics that gas or electricity could enable, by decorators such the Garretts.

The end of the nineteenth century saw the rise of the decorator as a distinct profession, with a literature that spoke to its aims and ambitions.[33] In *Suggestions*, the Garretts tried to clarify what was meant by the "misunderstood term 'decorator'":

> Until lately a house-decorator (to all except the extremely wealthy) has meant simply a man who hangs paper and knows mechanically how to paint wood. In his proper place he would fulfil the part which a dispenser does to a doctor, he should be able faithfully to follow directions, and honestly to carry out instructions ... But a decorator should mean some one who can do more than this; he should be able to design and arrange all the internal fittings of a house.[34]

Despite their use of the pronoun "he," the Garretts were ahead of their time in their comments directed to women. Compare their comments, for example, to Mrs Panton, another advisor to the decoration of the home, critiquing nineteen years later "the woman who demands to 'live her own life' and 'develop her soul' at the expense of the comfort of the household which she has undertaken to guide when she became the wife of the bread-winner."[35] Historians are generally agreed that the kinds of reform embedded in this and other kinds of "maternal feminism" were negligible in relation to some of the more controversial and radical feminism being advocated by a few women in this era. Indeed, some historians argue that women's so-called consumer power in the middle-class home simply cemented their gendered oppression by linking them ever more firmly to the chains of domesticity.[36] However, as Lynne Walker has argued, from the vantage point of the middle-class women themselves, women's increased involvement with design, as both consumers and as new professionals and businesswomen, nevertheless provided some early (and albeit tentative and non-revolutionary) pragmatic steps in the direction of increasing economic and political independence that would eventually bear fruit.[37] For, as Walker has persuasively argued, the Arts and Crafts movement prepared the ground, in some quarters at least, for the women's suffrage movement. This was due, in a large part, to the organizations that grew up through the social structures of the movement, modelled on the guild system, including the Arts and Crafts Exhibition Society, in which the Garretts participated in the 1888 iteration, after exhibiting a room "decorated and furnished by R. & A. Garrett" at the 1885 Exhibition of Women's Industries in Bristol.[38] Here they designed a complete interior, consisting of metalwork, including lighting.

Such exhibitions demonstrated the movement's new attitude to home decorating, and a shift away from previous designs, critiqued as dishonest for disguising the products and processes of manufacture. Objects that celebrated craftsmanship, materiality, and function were seen as conforming to Christian values and providing a good moral influence in domestic life. The Garretts clearly articulate the Arts and Crafts origins of their ideas: "Not even the most rabid 'Queen Annist' would deny to Pugin, and the other architects to whom the Gothic revival was due, the need of gratitude which their careful work merits."[39] Good, honest design was weighted with the idea of proper Britishness, and therefore the pressure increased on those tasked with choosing how to spend their money when building and decorating their homes.

Morality, Beauty, and Efficency: The Garretts' Suggestions for Lighting the Home

In the first chapter of *Suggestions*, the Garretts describe "Houses as they are," characterizing a London home as full of machine-made objects decorated to disguise their form, materiality, and function. This is followed by "Houses as they might be," in which the Garretts describe the same London house, but treated according to their Arts and Crafts principles, and in this we first see their Arts and Crafts ideas on the lighting of the home: "In the arrangement of a London house it is a matter of primary importance to avoid as much as possible darkening the rooms." The Garretts' advice included using colour that "reflects the light instead of absorbing it," and employing plate rather than coloured

4.1 "Hall Table and Chair," from *Suggestions for House Decoration in Painting, Woodwork and Furniture,* by Rhoda Garrett and Agnes Garrett

glass that "makes the most of the scanty supply of daylight with which Londoners are obliged to be content."[40] In their guidance to lighting the home, we can see the Garretts applying their Arts and Crafts principles to argue for a truth to materials and an honesty in design that reflected rather than disguised function and purpose.

In their discussion of the dining room in their ideal London house we find a gas-lit "chandelier":

4.2 "Dining-Room Chimney-Piece," from *Suggestions for House Decoration in Painting, Woodwork and Furniture*, by Rhoda Garrett and Agnes Garrett

So many really well-designed gas and candle pendants can now be bought wrought in polished brass and in forms consonant with the lines and curves that metal may legitimately take, that there is no excuse for the cast gaseliers coloured to imitate bronze that are found suspended over too many dining-room tables, and from which, it is true, light of a material kind can be obtained, but not that light which comes from the contemplation of every beautiful form whether it be fashioned in metal or in wood, as a humble article of domestic use, or as some greater work of art in painting or in sculpture. One of the great snares of the metal-workers of the present day is their aptitude to reproduce ecclesiastical forms in domestic work, and a purchaser must be constantly on his guard in choosing the metal fittings of his house or he will have a church corona over his dining table and an ecclesiastical scuttle for the reception of his secular coal. Even when the main outline of the work is not of an ecclesiastical character, Gothic crosses and battlements and finials will assuredly crop up and spoil the propriety of the design: for ornaments which would be appropriate enough in a church are quite out of place when used to adorn articles of domestic use.[41]

The Garretts see the interest in the ecclesiastical in lighting design as being linked to an attention to history and the preservation of a limited amount of buildings, and as "the only designs that had any pretensions to artistic merit ... executed solely for ecclesiastical purposes; while the forms and ornamentation of metal work for domestic use were designed by men who had probably never had the opportunity of studying the beautiful examples of ancient metallic art to be found ... in many old towns in this country." The conservation of old English design was one of the key motivations behind the Society for the Protection of Ancient Buildings, an organization founded by William Morris, Philip Webb, and Rhoda Garrett. Limited access to examples of domestic metalwork had led to "the manufacturers ... having little knowledge of the fitness of things and thus graft[ing] on their domestic work the details of construction and ornament which was designed solely for the furniture of Gothic churches."[42]

Beautiful lighting, for the Garretts, had to celebrate the very materiality of its construction and function, and the skill of the craftsperson. In their chapter on "The Drawing Room," referencing back to their comments on the "dining room gaselier," the Garretts advise that when it comes to "gas and candle-brackets, door furniture, and bell levers etc ... nothing better than polished brass can be used in the manufacture of these articles," with an emphasis on "simplicity of form and delicacy of workmanship."[43] Their focus in their design advice on "an old rule worth remembering which says that 'ornament, in the use of natural form, should be such as is not inconsistent with nature modelled by art,'" directly linked their guidance as to the aesthetics of lights to a wider concern for a philosophy of honesty, truth, and morality.[44] As I have previously discussed in my work on Philip Webb's Arts and Crafts designs at Standen, perhaps the Garretts were also helping the nervous housewife bring gas into her home by linking a new energy form with the aesthetics of the past, and designing a chandelier that echoed the candlelit ones of history?[45]

When *Suggestions* was published in 1877, gas was still the norm and so it is not surprising that this is the type of energy we see used in their illustrations and their home, and they chose to employ gas even into the 1890s.[46] This is evidenced by an interview for the *Women's Penny Paper* in 1888: "The bureau was cunningly ensconced in a corner by the fireplace, beside which projected a gas bracket holding a couple of opal-coloured globes, though the rest of the room appeared to be lighted by candles."[47] This preference for gaslight may have been influenced by the fact that Newson Garrett, Agnes's father and Rhoda's uncle, formed Aldeburgh Gas Light Company in 1856. Shares in the company were held by a large number of family members.[48] When Agnes won the tender to decorate Elizabeth Garrett Anderson's New Hospital for Women in 1892, each bed had its own light. There was a discussion during construction about the method of illumination. The cost of fitting for gas and electric was roughly the same, but the running costs for electricity were greater. Minutes record that "Mrs Anderson was ... getting some private interest to bear on the Midland Railway Co to see if they could supply the Hospital." It is likely that she approached James Beale, of Standen, to tap into the supply generated to run St Pancras Station, but this plan came to nothing and the hospital was lit by gas even though it was wired for electricity. The use of gas made it necessary to regularly clean and paint the wards each summer, and in 1892 this contract was awarded to Agnes. In 1896 she was commissioned to undertake the cleaning and repainting of the interiors, therefore, ironically, Agnes benefitted from the use of "dirty" gas in terms of future employment.[49]

To see how the Garretts' ideas on lighting were translated into actual designs, we can turn to the illustrations in *Suggestions*, as it has been argued convincingly that these were taken from real life and are views of the Garretts' own rooms at No. 2 Gower Street.[50] We can see the "old-fashioned bureau," the gas globes, and the rugs as described in the *Women's Penny Paper*, and they include furniture known to have been designed

4.3 "Drawing-Room Chimney-Piece," from *Suggestions for House Decoration in Painting, Woodwork and Furniture*, by Rhoda Garrett and Agnes Garrett

by the Garretts. The Garretts also took on a showroom or "warehouse" in March 1879 at 4 Morwell Street, and used it as the setting for at least one exhibition reported in the *Women's Penny Paper* in June 1887. The exhibition demonstrated a whole room setting, as was becoming the norm for the display of interior furnishing by the 1870s, encouraged by the use of period rooms in the South Kensington Museum. This allowed the Garretts to show their lighting designs alongside their designs for furniture, wallpaper, and textiles: "On four successive afternoons last week numbers of visitors availed themselves of an invitation to see this veritable model of what a sitting room should be ... some brass sconces were hung on the walls, most of them designed by Miss Garrett and all exquisitely worked in brass and copper by a real artist worker." This is one of the key pieces of the scant evidence that we have, alongside the illustrations in *Suggestions*, of their designs for metalwork and lighting, especially as very few of the identified objects have survived.[51] The "real artist-worker" has been identified as Alfred Shirley, credited in the Arts and Crafts Exhibition catalogue as sharing with Agnes the design of a brass and copper pendant and candle sconce, which he then made for the firm.[52] The aesthetics of such lighting was key when it came to persuading women to bring new energy technologies into their homes, hence the importance of looking at the history of energy through the lens of the history of design.

Mrs Haweis, the Arts and Crafts of Decoration, and the Housewife's Moral Challenge

The Arts and Crafts stance to the aesthetics of lighting was also adopted Mrs Mary Eliza Haweis (1848–1898) in her advice guide. Whereas the Garretts, writing in the 1870s, could only envisage gas, oil, and candle-light, Mrs Haweis turns to the challenge of "Lighting and Ventilation" in chapter 8 of her *The Art of Decoration*, and produces one of the earliest examples of a woman advising women on electric lighting in England that I have come across to date.[53] Given the date of publication, in 1881, at the very earliest period of electrification, and only one year after William Armstrong had first lit up Cragside using Joseph Swan's lightbulbs, it is unsurprising that Haweis starts her chapter on lighting with the statement, "Until the electric light is more manageable than it now is, there are but two ways of lighting rooms – gas or lamps and candles."[54]

The Art of Decoration, produced in 1881, four years after *Suggestions*, combined the history of art and design with advice to the nervous housewife, tasked with furnishing her home. In contrast to the Garretts, Mrs Haweis sees her task as an academic rather than a professional one. She identified as an author rather than a decorator and the first two sections of her book are mainly concerned with her version of the history of European art, as the precursor to her guidance on the furnishing of the modern home. Through her call to women to free themselves from the subjugation of decorators as part of her wider interest in the political enfranchisement of women, we can, however, closely align her to both the Garretts' ambitions, and our theme of ambivalence, selling guidance by highlighting the many ways in which a woman can make mistakes in her home. Her book begins with a thinly veiled warning to the genteel woman able to afford the services of a decorator: "Most people are now alive to the importance of beauty as a refining influence."[55] Mrs Haweis saw the design of the home as not just about aesthetics and function, but about spirituality and care. This echoes the claims of Augustus Pugin, John Ruskin, and William Morris, who proposed that art and design could alter the attitude of a society that they framed as broken after the Industrial Revolution. She states that "the appetite for artistic instruction is ... ravenous," but warns that, "the vacuum can be filled as easily as the purse can be emptied. Just now every shop bristles with the ready means: books, drawings, and *objets de vertu* from all countries are within everybody's reach, and all that is lacking is the cool power of choice."[56] Into this world of consumer goods, as described by Karl Marx in *Capital* (1867), such guides (both in terms of the books and their authors) became vital, to help avoid social *faux-pas*, and when we consider the challenges of the technologies of new processes of energy supply and the very real concerns of women about their potentially devastating impact on both their health and their beauty, the role of women as decorators, as advisors to women, and as shoppers becomes key.[57]

A call for the design of lights to be equally as moral and as reflective of beauty can be found in a 1904 article in *Art Journal* on "Electric Light and the Metal Crafts": "One of the distinguishing characteristics of old-world design is an inimitable blending of the utilitarian and the beautiful in the production of those everyday articles [here with direct reference to electric lights] which we, in these degenerate and restless days, with few exceptions, no longer think worthy of time or attention."[58] The article goes on to quote Ruskin's demand that "we should ornament construction, and not construct ornament."[59] This picks up precisely on Mrs Haweis's general exhortations on good design. Objects that celebrated craftsmanship, materiality, and function were seen as conforming to Christian values that domestic life should demonstrate.[60] No material should imitate something that it is not and furnishings that obeyed the criteria of beauty, it was argued, were bound to have a good moral influence. Robert Edis, for example, in his 1883 guide, *Our Homes and How to Make Them Healthy*, warned against the dangers of "dishonest design": "If you are content to teach a lie in your belongings, you can hardly wonder at petty deceits being practiced in other ways."[61]

Beauty and Fight against Falsehood

This focus on the idea of false or immoral furnishing is picked up right from the start in Mrs Haweis's book.[62] Her verbosity frames the same ethos as the Garretts', demanding that we lose "all the false 'embossed mouldings' ... recalling nothing, in their vacant misconstruction of classic types ... something that no genteel woman would want to bring into her home."[63] She also repeats Arts and Crafts ideas when she speaks about the designers and makers of the past: "They worked by hand where we work by machinery; and the difference between the one, which bears evidence of an individual mind, over the other, which is quite unintelligent, must be clear to all."[64] In her chapter on "Art Is for the People," Mrs Haweis speaks to the power of the purchaser: "Changes must emanate from the public ... the upholsterer cannot afford to be independent of the people – he must supply their demand ... educate the public that they may recognise what is good, whether in colour, shape, or construction. Educate the workman that he may be equal to the coming demand."[65] The 1880s saw the rise of the department store, of Heal's and Liberty and Harrods, and with this came the ability for the customer to choose, but with choice came inevitable concern about getting it wrong. And with the pressure on women to create a house that reflected their morality as well as their taste, no wonder the need for a guide did not fade away with the move from working with one designer or architect to buying from a range of shops and companies. Mrs Haweis's comments on the role of the decorator could almost be read as a call for women to exercise their own rights: "The province of a decorator ... is not to take your house out of your jurisdiction; he might as well control all your possessions and sell every-thing he did not personally covet. His province is to help you in that mechanical part which you cannot do yourself. He may guide you; he must not subjugate you."[66] Ironically, she does this in a book that is precisely aimed at telling the housewife what to do in her home. Women are exhorted to "liberate" themselves, but, as we can see, their new roles as both consumers and professional decorators raise a number of paradoxes in terms of agency and emancipation.

A Woman's Guide to Lighting the Home

Mrs Haweis's chapter on "Lighting and Ventilation" is particularly interesting for our subject of women and energy, as it ultimately comes out on the side of electricity as the best form of lighting for the home at a very early point in the history of its domestic use.[67] Despite this pioneering opening statement, she does go on to focus her discussions on gas, although like other female commentators, she points to equally challenging issues with this form of lighting, saying that although it was "the cheapest, and the least trouble … it is the most destructive to furniture and pictures, the least healthy and the least becoming."[68] Mrs Haweis, like many of the decorators and suppliers of lighting at this time, considered the very oldest of lighting types as well as the newest, declaring that "Lamps are the next best, if they can be induced not to smell; wax candles are the best of all, if they can be warranted not to bow."[69]

She is focused throughout the book on the idea of the natural and nature, and this is as much the case with lighting: "The main light ought to be concentrated as much as possible in one spot. This is nearest to a natural effect for the sun is never in two or three places at once, and will be found becoming to faces and the folds of dresses."[70] Again, we can see that she is addressing her comments to women, and the concern for appearance under different lighting types is key: "It should always be remembered that faces look best (if we may venture to disagree with Queen Elizabeth) with their natural shadows, which give that 'drawing' to them always missed on the stage when the footlights glare up from below."[71] Having discussed the right type of light, Mrs Haweis goes on to think about "Lamp-Forms," and in this she echoes the Arts and Crafts advice that we can also see in the pages of the Art Journal: "May I remind readers that a candlestick, lamp or any other support, ought to be a pretty and consistent object."[72]

Conclusions

By 1900, efficient lighting remained a luxury in England, and so Mrs Haweis, in her recommendations for electric lighting in 1881, was very much ahead of the times:[73] "When the electric light comes into common use, the problem how to light adequately a large room without heating it will be solved. I have seen the picture gallery at the Fitzwilliam Museum, at Cambridge, successfully lighted by electricity subdued by a tinted globe; Lord Salisbury has introduced it at Hatfield."[74] She recognized that, for women particularly, new lighting types would alter the appearance of the home, and of those who spent most of their time in it, the women of the house. Therefore they were the ones most concerned not so much by the technologies, but by the aesthetics of change. In the battle between the gas and electricity companies, women were key in terms of convincing the consumer as to which energy source to plump for in the home. In the year that Mrs Haweis published her book, the House of Commons was lit by incandescent lamps,[75] the first great exhibition of electricity was held in Paris[76] and the electrical company Crompton installed a thousand Swan lamps to illuminate the Savoy Theatre.[77] But, as there was still no public supply of electricity, and current had to be provided from expensive generators installed on individual premises, these innovations attracted widespread popular attention but did not move as quickly into the home as the burgeoning group of electrical companies would have liked.

Writing this as we commemorate 100 years since woman's suffrage in England, it seems appropriate to look to the Garretts and Mrs Haweis as women who pioneered women's engagement with home lighting and decoration. But in doing so we need also to reconsider the ambivalent role of women as they took on the role of professional advisor to other women in the home while promoting universal suffrage. A number of scholars have cited mediation as a key concept when studying technological artifacts from social and historical perspectives.[78] Rachel Plotnick, in "At the Interface: The Case of the Electric Push Button, 1880–1923," wonderfully explores the role of the electronic mediator, the actual push button that one would need to push to turn the lights on and off, from its invention in the 1880s until its wholesale adoption by the 1920s.[79] Perhaps the answer to the question of ambivalence is to see the Garretts and Mrs Haweis through this lens of electrical mediation, not in terms of the actual technology, but in terms of those advising women on how to integrate it into their homes successfully. The term "mediation," when used in human resources practice for example, suggests a process of support and care. Can this interpretation offer us a way of understanding or framing the ambivalent nature of the first women decorators – guidance as care rather than hard-faced consumer practice? As such, we can see our first women guides to lighting the home as supportive, politically engaged, and professionally innovative.

Notes

1   See Langland, *Nobody's Angels*, for an excellent discussion of the changing expectations of middle-class women in the home in the nineteenth century.

2   Haweis, *The Art of Decoration*, 2.

3   "The 'Angel in the House' … the presiding hearth angel of Victorian social myth, actually performed a more significant and extensive economic and political function than is usually perceived. Prevailing ideology held the house as haven, a private sphere opposed to the public, commercial sphere.

In fact, the house and its mistress served as a significant adjunct to a man's commercial endeavours. Whereas men earned the money, women had the important task of managing those funds toward the acquisition of social and political status." Langland, *Nobody's Angels*, 8. See also, Davidoff and Hall, *Family Fortunes*.

4   B. Smith, "Havens No More," 100.

5   See, for example, Ehrenreich and English, *For Her Own Good*.

6   When *The Art of Decoration* was first published in 1881, Mary Eliza Haweis was listed on the cover and frontispiece as "Mrs H.R. Haweis," following the Victorian convention of women authors being named after their husbands. I have therefore used the title "Mrs Haweis" throughout this chapter, and where citing other women authors on the home of the same period, have listed them according to how their names first appeared in their publications. This demonstrates the Victorian convention on the presentation of married women-author's names, even when they were writing about emancipation. Agnes and Rhoda Garrett, in contrast, as unmarried women, published under their full maiden names. While using "Mrs" in order to recognize the gendered history of this practice, I do not want to suggest that I am supporting this outdated convention. Instead this is a perfect example of the paradoxes and problems of women's identities that we explore in our book.

7   Garrett and Garrett, *Suggestions for House Decoration*, 28.

8   C. Edwards, "Establishing Stability," 49.

9   Christopher Dresser, in *Studies in Design*, pointed out that "the decoration of a room is as much bound by laws and by knowledge as the treatment of a disease," 39.

10 Garrett and Garrett, *Suggestions for House Decoration*, 9–10.

11 See, for example, Joy Parr on male salesmen and their bullying tactics when advising women on their electrical purchases in 1940s Canada: "Shopping for a Good Stove," 75–97.

12 See SPAB Blog, "Women in Conservation: House of Garrett."

13 Ibid.

14 Crawford, "Spirited Women of Gower Street," 3.

15 Ibid., 3–4.

16 Here we see how many of these women authors were published under their husband's name. Emma Ferry pointed out the challenge of gendered conventions for styling married women's names, particularly in her work on Lucy Faulkner, Mrs Orrinsmith. Here, while commenting on how much less is known about Lucy Faulkner in comparison to her sister, the Arts and Crafts designer Kate Faulkner, Ferry points out that "a contributing factor to her comparative obscurity is simply that she married and changed her name" ("The Other Miss Faulkner," 5). She quotes Deborah Cherry's work on the

difficulties faced by women in "the making of an author name": "Those who married had to negotiate a change of family name and either re-establish their career with a second or sometimes third name" (Cherry, *Beyond the Frame*, 157). Ironically, given some of the comments in her book that seem to suggest a more old-fashioned view of a woman's place in the home, Jane Ellen Panton published her book under the name J.E. Panton, using her husband's surname but her own initials, and without the prefix "Mrs."

17 "Surprisingly, very little is written about the Garrett cousins. The biographical file on Rhoda Garrett at the RIBA library simply contains a photocopy of her obituary from *The Builder*." Ferry, "Decorators May Be Compared to Doctors," 17. As a result they have often been the victims of repeated and misleading errors that frame their work through that of their male contemporaries. For example, Asa Briggs in *Victorian Things* claims that *Suggestions* was written in collaboration with Owen Jones, who had died in 1874, before the book was published (ibid.). Elizabeth Crawford has produced the most significant body of work on the Garrett family, including her monograph *Enterprising Women*.

18 "The Glasgow International Exhibition: Industrial Art: Women's Arts and Industries," *Art Journal*, December 1888, 6.

19 In the second half of the nineteenth century, feminist concerns about the employment of middle-class women and their entry into the professions were linked to the specific profession of architecture and the campaign for married women's rights, culminating in the Married Women's Property Acts in 1870 and 1882. Before these Acts were passed, under common law, married women's property, earnings, and inheritances belonged to their husbands. Architects were often sued in connection to their contracts, and since, under common law, married women were not allowed to make contracts or be sued in their own right, they were precluded from many of the professional responsibilities of architecture. Therefore, "as long as women were the virtual property of their husbands, they did not, and could not, act in a professional capacity as the designers of property – property cannot design property." The removal of this impediment by the Act of 1882, therefore, had particular importance for the entry of women into the architectural profession. In the early 1890s, architects such as Richard Norman Shaw and William Lethaby argued that architecture was an art, not a business, and between the census of 1887 and that of 1891, the classification and status of the architect altered from "Industrial Class" to "Professional Class" under "Artist." If architecture was an art, and art was an area deemed appropriate for women, then architecture became a suitable profession for women. In 1898, Ethel Mary Charles became the first woman to enter the Royal Institute of British Architects, sixty years after its establishment. Walker, "Women Architects," 96.

20 "Glasgow International Exhibition," 6.

21 Forty, *Objects of Desire*, 105.
22 Conway, *Travels in South Kensington*, 169, quoted in Crawford, *Enterprising Women*, 170.
23 Conway, in his 1904 *Autobiography* (vol. 1., published by Cassell and Co.), 403, quoted in Ferry, "Decorators May Be Compared to Doctors," 18.
24 Conway, *Travels in Kensington*, 169. Ferry suggests this may be a reference to Girton College, University of Cambridge, but believes that it is more likely to refer to Newnham College. "Decorators May Be Compared to Doctors," 19. In her footnote she states that "Basil Champneys, who shared an office with J.M. Brydon and who must have known the Garrett cousins, was the college architect for Newnham College from 1874 until 1910. It is very tempting to imagine that Rhoda and Agnes Garrett may have decorated rooms there in the "Queen Anne" manner. Moreover, given that Millicent Fawcett, her husband and daughter Philippa had long and close links with Newnham, it seems highly likely that the Garretts were involved" (31, n45).
25 Crawford, *Enterprising Women*, 170–1.
26 Garrett and Garrett, *Suggestions for House Decoration*, 2, 5.
27 Cooper, N., *The Opulent Eye*, 8. There were twelve volumes in all in the Art at Home series, including Mrs Orrinsmith's *The Drawing-Room* (1877), Mrs Loftie's *The Dining Room* (1877, dated 1878), and Lady Barker's *The Bedroom and the Boudoir* (1878). Emma Ferry has written in detail about each of the volumes. See, particularly, "Information for the Ignorant and Aid for the Advancing," "The Other Miss Faulkner," and "Any Lady Can Do This without Much Trouble."
28 Ferry, "Decorators May Be Compared to Doctors," 23.
29 Rev. W.J. Loftie, author of *A Plea for Art in the Home* (Macmillan and Co., 1876), in correspondence with his publisher, "Letter to Macmillan," 11 March 1876, cited in Ferry, "Decorators May Be Compared to Doctors," 23.
30 Ibid., 24.
31 Garrett and Garrett, *Suggestions for House Decoration*, 6.
32 See Harrison Moore, "Designing Energy Use in a Rural Setting"; Gooday and Harrison Moore, "True Ornament?"; and Harrison Moore and Gooday, "Decorative Electricity."
33 Edwards and Hart, eds., *Rethinking the Interior*, 5–6.
34 Garrett and Garrett, *Suggestions for House Decoration*, 5–6. The Garretts were positioning themselves as professional decorators for hire, but, as Mrs Panton reminds us in 1896, nineteen years after the publication of *Suggestions*, it was still difficult for a middle-class woman to have a professional job, despite the fact that she was writing a guide to decorating the suburban home: "Where a house fails to be a real success, be sure it is because some detail has been forgotten … and here I am sure, is a great opening for an artistic woman with a keen eye for effect – if only she could persuade some of our larger

upholsterers to employ her – for such a woman is badly wanted to go round every newly furnished house, and note down as she goes the small things which will invariably escape the best male eye in the world, and even the eye of the best decorator." Panton, *Suburban Residences*, 62. This shows how ahead of the times the Garretts were, writing and advising in 1877 as women decorators, given that this, for Mrs Panton, was still considered an unlikely profession for a woman.

35 Panton, *Suburban Residences*, 70–1.
36 See, for example, Callen, "Sexual Division of Labour in the Arts and Crafts Movement"; and Parker and Pollock, *Old Mistresses*, 99.
37 Walker, "The Arts and Crafts Alternative."
38 Ferry, "Decorators May Be Compared to Doctors," 19. She cites the *Women's Penny Paper*, 18 January 1890, discussing Agnes exhibiting at Bristol, "where only women's work was admitted."
39 Garrett and Garret, *Suggestions for House Decoration*, 2.
40 Ibid., 38.
41 Ibid., 53.
42 Ibid., 52–3.
43 Ibid., 66.
44 Ibid., 81.
45 See Harrison Moore and Gooday, "Decorative Electricity."
46 See Gooday, *Domesticating Electricity*.
47 Interview conducted at Gower Street for the *Women's Penny Paper*, 3 November 1888, quoted in Crawford, *Enterprising Women*, 194.
48 Crawford, *Enterprising Women*, 20–1.
49 Ibid., 66–8.
50 See ibid., 177.
51 Crawford states that a "family member" still holds a collection of brass door furniture. Ibid., 186.
52 Ibid., 186–7. Crawford reports that Colin Anderson, grandson of Elizabeth Garrett, mentioned that the Garretts certainly designed "finger plates for doors, tiles for fireplaces, light fittings and banisters" (189).
53 Mrs Gordon published her famous book, *Decorative Electricity*, so carefully considered in Gooday's work, in 1891, ten years later.
54 Haweis, *The Art of Decoration*, 350.
55 Ibid., 2.
56 Ibid.
57 On this point, see Gooday, *Domesticating Electricity*, beautifully illustrated by a reproduction of a cartoon in *Punch* magazine. The idea that domestic décor was an expression of a woman's personal character was further linked with the idea of the home as a reflection of and influence on a person's morality, and with the importance of cleanliness, a concept that was extremely

prevalent by the later nineteenth century. The Garretts and Mrs Haweis both make suggestions that link health and the decoration of the home, and while I do not have space here to examine these in detail, there is an important link made between energy decisions and health in all three women's work. There is a growing and rich secondary literature on the history of the relationship between health, sanitation, and the home that will be of even greater interest post-COVID-19. See, for example, Tomes, *The Gospel of Germs*; Haley, *The Healthy Body and Victorian Culture*; Kiechle, *Smell Detectives*; and Nash, *Inescapable Ecologies*.

58 "Electric Light and the Metal Crafts," *Art Journal* (October 1904): 321–8.

59 Ibid.

60 Forty, *Objects of Desire*, 112.

61 Edis, "Internal Decoration," 356.

62 Haweis, *The Art of Decoration*, 2.

63 Ibid., 3.

64 Ibid., 18.

65 Ibid., 21.

66 Ibid., 29–30.

67 Ibid., 350.

68 Ibid.

69 Ibid.

70 Ibid., 350–1.

71 Ibid., 351–2.

72 Ibid., 353.

73 "There was a larger demand for lighting in Britain than in the less urbanised and less well-lit USA, but the lion's share of the market in this country went to the gas industry." Hannah, *Electricity before Nationalisation*, 9. Byatt, in "The British Electrical Industry," 56–7, calculated that in 1890 to 1891 the gas companies' share of the gas and electric market was 68 per cent in the US, and almost 100 per cent in England and Wales.

74 Haweis, *The Art of Decoration*, 353.

75 Note from Gooday, *Domesticating Electricity*, 241. "Much interest attended the introduction of the electric arc light into the House of Commons in 1881 but it was removed by January 1882 and 'illuminated by the customary means,' an embarrassed euphemism for coal gas lighting." *Electrical Review* 10 (1892), 2.

76 The first of the great electrical exhibitions was the Paris Electrical Exhibition in Autumn 1881 and was reported on almost daily in *The Times* of London. This was followed in Spring 1882 by the International Electrical Exhibition held in Crystal Palace in London from Jan to June 1882, see Gooday, *Domesticating Electricity*, 93–8, and also *The Times*, 23 March 1882. "Those who have visited the Electric exhibition at the Crystal Palace will have seen

for themselves beautiful miniature lights, admirably adapted for domestic use, with glare and flickering subdued," "The Electric Lighting Bill," *Times*, 14 April 1882, 9C. The Electric Lighting Act was passed in August 1882, legislation designed to licence the public electricity supply.

77 Hannah, *Electricity before Nationalisation*, 4.

78 See, for example, Ihde, *Bodies in Technology*; Luber and Kingery, *History from Things*; and Verbeeck, *What Things Do*.

79 Plotnick, "At the Interface," 815–45. More recently, Sandy Isenstadt has provided a chapter on the electric push button, "At the Flip of a Switch," in *Electric Light*, 24–65.

5

Women in Energy Engineering:
Changing Roles and Gender Contexts in Britain, 1890–1934

Graeme Gooday

The Electrical Association for Women makes this contribution to the education of the woman public, in an endeavour to bridge the gap between the electro-technical and the domestic worlds.

Margaret Moir, EAW president, and Caroline Haslett, EAW director;
12 March 1934[1]

Introduction

What sort of "gap" was there between the "electro-technical" and domestic worlds of interwar Britain? And why did the UK's Electrical Association for Women (EAW) ask the "woman public" to read its *Electrical Handbook for Women* to close this gap? That *Handbook*, published in 1934 to mark the EAW's tenth anniversary, bears gendered narratives of expertise and responsibility concerning energy production and consumption. A clue to the agency of the "woman public" in the troubled relationship between industrial supply and domestic consumption is evident in the *Handbook*'s front cover. Designed by fashionable Art Deco illustrator Ethel "Bip" Pares, it depicts a pair of stylized, long-fingered "feminine" hands cradling the armature of a dynamo.[2] While in part capturing the traditional gendered prerogative of manual "care," the illustration also implies a gendered dynamism. It was evidently labour by women's hands using domestic electrical technology that would make the dynamo turn, thereby generating a supply of electrical current. The success of the electrical supply industry required a clear demand from the "woman public" for domestic electricity.[3]

There were indeed various respects in which women's roles were *necessary* for the dynamo to turn – that is, women were not merely acting as passive recipients of this energy supply, as conventionally construed

THE
ELECTRICAL
HANDBOOK
FOR
WOMEN

5.1   Front cover for the Electrical Association for
Women's *Electrical Handbook for Women* (1934), edited by
Caroline Haslett

in energy history. As I will show, a small number of women such as
Margaret Partridge were involved in the design and manufacture of
electrical power technology and/or the management of electrical supply
systems. More numerous were the hundreds and eventually thousands
of EAW lecturers and demonstrators on electrical housecraft who were
aiming to maximize the number of informed consumers of domestic
electricity among the millions of householders deciding whether to
use electricity or "town gas" to power and light their homes. In many
ways the EAW staff and these householders were arguably the most
important: to echo a theme from my book *Domesticating Electricity*
(2008), covering an earlier period, if female homemakers could not
be persuaded that they needed to abandon older cherished systems
of energy (coal, wood, gas, and paraffin) and adopt electricity instead,
there would be no demand to summon the dynamos into turning.[4]
The tacitly acknowledged challenge was, I suggest, the *skepticism*
among the domestic female public as to the merits of electric energy

supply. Only a body led by women – namely the EAW – could muster the authority to change women's preferences to electricity and thus their energy consumption practices.

The numerous editions of the *Electrical Handbook for Women*, edited by Caroline Haslett, were aimed primarily at supporting the national network of EAW staff: "the teacher, the demonstrator, the lecturer and every other woman who has any interest, whether professional or academic, public or social," in the "correct use and control" of electricity in the home. This "correct" approach would be inculcated by supplying these readers with the required "technical background" on the production, transmission, control, and "use of electricity."[5] In contrast to the usage of older household energy supplies, Haslett envisioned the ideal female consumer of electricity as a skilled practitioner of a scientific craft, formally qualified with EAW credentials. The *Handbooks*'s diverse array of male and female contributors (discussed in detail later) epitomized the matrix of gendered relationships required to harmonize consumers' household demand to industrial supply. This intepretation thus challenges teleological portraits of the electricity industry as inevitably bound to succeed on economic grounds alone.[6]

The first of the two *Handbook* prefaces was co-written by Haslett, a First World War engineer and boiler maker, and the *soi-disant* "engineer-by-marriage" Margaret Moir. They had been two of the broad group in 1919 that launched the UK's Women's Engineering Society (WES), which, as several historians have noted,[7] spawned (not uncontroversially) the EAW as a side project five years later. While the EAW's creation of a large cadre of women lecturers and trainers in "domestic electrical engineering"[8] enabled the commercial success of the electrical industry, I would emphasize that such EAW career paths were created by Haslett and others to compensate for the enormous difficulty WES had encountered in creating careers for women in engineering during the challenging economic conditions of the mid-1920s – even for women who had been employed as electrical engineers during the First World War. Conversely in the *Handbook*'s second preface, we can see (in microcosm) how much the "woman public" was in turn needed by the electricity industry to succeed. There Sir John Snell, chair of the UK's Electricity Commission, praised Haslett and Moir's efforts to reduce housework drudgery, while confessing his own professional reliance on the expertise of his "helpmeet" spouse – both of which were needed to win over more householders to electricity.[9]

This male engineer's reliance upon a collaborative spouse is a tradition that went back half a century to a mode of marital endeavour unacknowledged in traditional histories of engineering. In the next section I show how such work by Alice Gordon and Katharine, Lady Parsons both broadens our understanding of women's role in electrical supply and accounts for the prevalence of male engineers' spouses in the early management of the WES. In the following section I explore how the First World War created nonfamilial routes for women such as Haslett to enter (energy) engineering. Finally I explore how Haslett and others in WES created the EAW and shaped its activities in ways that produced the *Electrical Handbook for Women*.

In so doing I present an alternative to Carroll Pursell's interpretation that the EAW was primarily created to overcome female householders' alleged ignorance of electricity – a "deficit" account that overlooks the way that such householders' skepticism about the merits of electricity were grounded more broadly in issues of (un)trustworthiness, caution, and economy rather than mere unfamiliarity.[10] Instead I build upon Suzette Worden's analysis of women's agency and the discretion of women in the EAW's story, especially in the contrasting feminist agendas of WES and the EAW: the former nurtured opportunities for women in engineering and the latter secured opportunities for women to escape domestic labour.[11] Both forms of female agency are vital to our understanding of how the consumption "gap" between domestic life and electrical industry was managed by both sides.

The Energetic Engineering Spouse

Energy historians a generation ago typically treated women encountering the new domestic electrical technologies of the twentieth century as passive consumers, even arguably as victims of patriarchal indifference to any desired autonomy to shape these new technologies to their own preferences. In *More Work for Mother*, Ruth Schwartz Cowan's classic study of how new utilities were developed for the US home, she shows how, of all the potentially more liberating ways that could have been chosen to make the homemakers' burden lighter, the one that emerged involved women still being tied to housework by machines that perhaps reduced their drudgery but not their overall labour. Schwartz Cowan's view is that middle-class women who had lost their servants by the 1920s were effectively proletarianized into machine minders for household equipment. This result was not one that matched the advertising rhetoric of the electrical industry, which promised that electricity would bring freedom from back-breaking labour.[12] Perhaps not coincidentally, Schwartz Cowan's view of women as powerless to shape the new electrical energy supply industry was reinforced by Thomas P. Hughes's

contemporaneous *Networks of Power*. As a former engineer, Hughes
portrayed that industry as exclusively managed by male engineers on a
strictly system-building engineering agenda, without reference to any
householders' expertise, needs, or interests.[13] Thus we have inherited a
"received view" that said nothing about the small number of women
who were actively involved in managing the centralized generation and
distribution of energy.

Further to that omission, unlike the traditions of industrial history
and agricultural history wherein women's labour within the family
setting is well-documented,[14] almost nothing has been reported of
spousal engineering labour.[15] More instead has been written about
women's efforts across Europe and the usa from the late nineteenth
century to become members of the engineering profession, independently of any traditional family support network.[16] Nevertheless in the
pre-corporate era, it was clearly common for male business practitioners
to involve their spouses routinely as de facto collaborators in their
business enterprises. As I argued in *Domesticating Electricity*, "electrical
engineering wives" played an important role in supporting the family
business to survive (and possibly) thrive by advising female householders
on how to domesticate electricity to their purposes. To extend that
discussion, I analyse next how two such engineering spouses – Alice
Gordon and Lady Parsons – actively assisted in the energy supply
business by developing and managing machinery, and the associated
managing workforces.

In 1891, "Mrs J.E.H. Gordon" produced a best-selling book, *Decorative
Electricity*, in collaboration with her electrical engineering spouse, James
Edward Henry Gordon.[17] That work closes with Alice's chapter "Some
Personal Experiences." This an unprecedentedly candid account of the
challenges facing a woman obliged to become an "engineer-by-marriage."
Alice represents herself as fully immersed in the stressful, arduous, and
risky work that James Gordon undertook as a professional consulting
engineer. Far more than a passive "helpmeet," she gave emotional,
managerial, translational, and practical support as deputy in his everyday practice of trialling new power-generating machinery and then
of installing it. Most evident, and not discussed in any other history of
electrical energy supply, is the emotional context of the shared pain
of failed technological experiments, unreliable workers, demands for
lawyers' fees, and much more besides. In this rare, intimate close-up
spousal view of engineering, she highlighted in fact how her endeavours
were "types of those of many others working in the same field" – an
allusion to the many other engineering spouses who she knew with
similar responsibilities.[18]

The public who through long winter evenings, and longer London fogs, sit reading by the cool and steady light of their electric lamps, but who are most indignant if by any chance it flickers or fails them, do not realize how intense the struggle has been for those pioneers of electric lighting who have toiled so hard and incessantly to surprise yet one more of Nature's secrets; and "To wring from iron those drops of golden light." Perhaps no one but an electric engineer's wife can truly judge of what that struggle has been.[19]

It would take "a long book," she wrote, to tell of "all the hopes and fears and disappointments" of her joint work with James in the 1880s, with alternate current dynamos that exploded into tiny pieces after just starting to work."[20] But as she noted, once such a machine could supply current without self-destruction, it could be installed on a permanent basis; and such is how the Gordons succeeded in lighting London's Paddington railway station in 1885 to 1886. Her knowledge of that installation was revealingly comprehensive:

Engines and boilers of nearly 2,000 horse-power in all were erected, and three dynamos, weighing forty-four tons each, were designed and built for the purpose. Eleven miles of iron troughing were buried between the lines, and in this some 500 miles of insulated wire were laid.[21]

Notwithstanding severe skepticism and doubts from male "experts" on the viability of this installation, it operated at Paddington station continuously from 6 p.m. on 21 April 1886 up to the time she wrote *Decorative Electricity*. Such was her expertise, that while James was away on consulting business abroad, Alice took on the task of overseeing the supply system's operations, sending him news by telegram. Those three months, she reported, were "my very hardest personal experience," visiting the station regularly to seek to allay James's "keenest anxiety" about both the machinery and impending lawsuits about its noise disturbance. During his absence, unsympathetic commentators sought to frighten Alice by describing the Paddington works as "constructed with seven volcanoes, any one of which might explode at any moment."[22] Nevertheless, the unexploded Paddington operation, and the less fraught sibling stations that the Gordon team worked on elsewhere in London, effectively put their expertise in demand among those seeking to supply electricity.

Tacitly underpinning the very writing of *Decorative Electricity*, however, was the lack of demand for domestic electrical supply in London.[23] Alice's expertise in how to develop elegant indirect forms of electric lighting was evidently essential to drum up custom for what many women experienced as an unpleasantly glaring illuminant that was more expensive, less reliable, and uglier than its gas, candle, and paraffin counterparts.[24] The Gordons' lack of success in that endeavour, however, resulted in such financial difficulties that, upon James's sudden accidental death in 1893, Alice was obliged to withdraw from engineering activity and became an independent writer, never to participate in energy supply again.[25]

A rather greater success came for Katharine, Lady Parsons (1859–1933) working with the Honorable Charles Parsons in their development of the steam turbine in the 1890s. By the interwar period this technology became the high-efficiency replacement for the traditional condensing steam engine. Starting life in East Yorkshire in 1859, Katharine Bethell met the Anglo-Irish aristocrat Charles Parsons while he was working as an engineer in Leeds in 1882 to 1883. She soon became an "engineer-by-marriage," showing her support for his experimental work by going, shivering in mid-winter, with Parsons and his mechanic at 7 a.m. to test the operation of a self-propelling torpedo on Roundhay Lake. When they moved to Tyneside, their marriage and creative engineering work was further entwined as they produced the first version of the steam turbine in 1884. When she was not working on the Conservative campaign for women's suffrage, she would often be present at the Gateshead works in administrative tasks for the Parsons Company and managing the workforce – which Charles was not skilled at handling effectively. Moreover, he operated a workshop at their home where much of the later development of the steam turbine took place – presumably away from the eyes of potentially untrustworthy employees, but also, it seems, so as to have Katharine close to hand while he worked into the night.[26]

The early stages of the steam turbine's development brought experiences akin to those of the Gordons, with a huge challenge of domesticating this innovation in reliability. While Lady Parsons was the "possessor of a remarkable character" and "almost fiercely independent" in every other matter, one obituarist for an engineering journal remarked that, when it came to innovative engineering work:

She was always at [Charles's] side, always there to help him when he needed her, always supporting him with her really powerful mind and ready tact, and perfect understanding, but quietly and unobtrusively from behind the scenes ... All those who worked or played with Sir Charles were struck by [Katharine's] constancy as his mainstay and support.[27]

Charles Parsons's biographer, writing with the direct advice of Lady Katharine and her daughter, Rachel, recorded that Katharine and their children (Rachel, born 1885, and George, born 1886) had been on board at the first trials of the steam turbine–driven *Turbinia*, circa 1894 – the entire family being drenched by the bow wave as the ship accelerated to an unprecedented twenty-eight knots.[28] As the hyper-efficient steam turbine eventually came to be used in both ships and electrical power stations in the ensuing decades, Katharine both shared in the reputational success of the Parsons invention and acquired her own independent authority as an engineer. This was particularly highlighted during the First World War as she managed numerous munitions factories in the North East, thereby earning an Honorary Fellowship of the North East Coast Institution of Engineers and Shipbuilders in 1918.[29]

It was the authority that she secured as an engineer in her own right that gave her a platform from which to promote the longer-term employment of women as engineering professionals, building upon the substantial role of women taking up engineering roles in the First World War. In a widely publicized speech in 1919, Lady Parsons argued robustly that, irrespective of the return of male engineers from the battle-front to reclaim their roles, women should continue to be employed in engineering of all kinds in peacetime, too. Indeed, in collaboration with her daughter, Rachel (who had studied mechanical engineering at the University of Cambridge), and others, she helped to found both the WES in 1919 and the Atalanta Company the following year, which employed only female engineers in making such things as oil burners.[30]

Having vigorously supported her spouse, Charles Parsons, in launching (literally) the steam turbine enterprise, Katharine secured his support for the WES project. In relation to this, the (female) obituarist for the North East Coast Institution of Engineers and Shipbuilders reported in 1933 that she had a "love of business organization and leadership." This, the obituarist opined, was "proved over and over again by the devotion of those who served with and for her – even in the humblest capacity."[31] Yet this was not quite the case when Lady Parsons hired the energetic Caroline Haslett as secretary of the WES in February 1919:[32] six years later, Haslett had redirected her numerous female supporters to the less radical enterprise of the Women's Electrical Association. It is to that topic I now turn.

Caroline Haslett: Dynamo of the Women's Electrical Association

Having become a suffragette in 1913, Caroline Haslett ended her activism at the outbreak of war in 1914 to take a secretarial course. She joined the Cochran Boiler Company from 1914 to 1918 as a junior clerk, drawing up quotations and specifications for boilers, the only woman employed to do so. So successful was she in this endeavour that she ended up managing the company's London office by 1918, and supplying boilers to the War Office. Even with the large-scale replacement of male engineers by female engineers during the war, she later wryly recalled her confident encounter with a War Office official who responded to her personal visit: "What are we coming to if a wisp of a girl can talk about boilers?" She soon moved to Scotland to work in the Arran office to learn the practical details of boiler making, as she had requested, recalling, "As I got into the works, I knew this was my world" – she designed and made boilers, and even sold some directly using the genderless name "C. Haslett" in all correspondence. Although many women's jobs in engineering (and elsewhere) disappeared after the war ended, the Cochran Company so valued Caroline's work that she was kept in employment.[33]

However, in January 1919 one of her (anti-feminist) male colleagues at Cochran's alerted her to an advertisement in engineering journals: "Required: Lady with some experience in Engineering Works as Organizing Secretary for a Women's Engineering Society." Haslett applied only reluctantly, but turned out to have a winning combination of skills both in running an engineering department and writing shorthand. Very surprised to be appointed, Haslett began work in spring 1919, setting up the twofold aims of the WES articulated at the Society's public launch on 23 June of that year: to promote the "study and practice of engineering among women" and to enable such "technical women" to discuss shared interests, opportunities for training and employment, and associated publications.[34] Haslett's work also included drumming up membership and editing the Society's commercial publication: the *Woman Engineer*, launched in 1920 to publicize the growing number of women active in engineering.

One piece from the 1920 issue of the *Woman Engineer* on the taming of dynamo technologies was by one of the WES's early members, the University of London mathematics graduate Margaret Partridge.[35] She was an independent electrical engineer who not only set up and operated independent power stations round the UK, but later wrote the entirety of the *Electrical Handbook*'s first section of six chapters on "General Principles of Electricity": a clear illustration of the authority that could be accomplished by a female engineer of energy supply.[36] Partridge initially tried her hand at teaching after graduation in 1914 but moved to wartime munition engineering work in 1915, as did hundreds of others in her situation. She thus gained extensive experience in electrical power technology, from which she was able to set up her own consultancy operation as "M. Partridge and Co., Domestic Engineers" (again note the use of an ungendered name). Given the challenges to women's continued employment in engineering – as raised by Lady Parsons – serving as a freelance, self-employed consultant was one of the few modes of professional engineering work open to women, especially in the economic recessions of the interwar period. Partridge's company was soon successful enough, however, to be able to advertise that it was hiring female electrical specialists: the *Woman Engineer* called for "Women for Women's Work" – providing lighting and electric power in rural locations. Partridge's first electrical power installations were completed in the Southwest of England in 1925 and 1926 with shareholder support from Lady Parsons and Lady Shelley-Rolls. As Partridge wrote to her by-now close friend and EAW ally Caroline Haslett from the new power station in Bampton, Devon, "My dear – for sheer exciting experience give me a town to light."[37]

While Partridge's career grew upon such successes, writing regularly on electrical machinery for the *Woman Engineer*, Haslett's experience as secretary became more challenging as the recruitment of new WES members slowed to a near halt, since Haslett's salary was paid entirely from the membership fees of the women (and a few men) who joined the WES, plus a discretionary sum invested from Lady Parsons's personal wealth. The recession of 1922 dampened WES initiatives, and the Society was only rescued from near bankruptcy by generous donations from philanthropic women such as Lady Beilby.[38] WES's awkward reliance on such fortuitous funding soon prompted Haslett to seek alternative, domestically centred careers for women with technical expertise in electricity. Specifically, in 1924 she asked WES members what they felt had been the most important engineering initiative in home efficiency: the largest vote was for a dishwasher machine, and this prompted discussion on how best to develop this device to enhance household electrical consumption.

A further important encounter that shaped Haslett's thinking in this direction was with the American time and motion specialist Dr Lillian Gilbreth, whom Haslett met at the first World Power conference, held at London in 1924.[39] This encounter made Haslett particularly amenable to a paper on women's domestic uses of electricity offered to her in the summer of 1924 by a WES member, Mrs M.L. Matthews (who had previously used it unsuccessfully to try to gain membership to the Institution of Electrical Engineers). Having come independently to time and motion analysis from her encounter with a rural haymaker during the war, Matthews contended in her paper that too many women overlooked the fact that "the thrift of one's energies is often more important than thrift of money. It is by this form of thrift that electricity is going to help women." The upshot of this was that Matthews suggested the formation of an Electrical Association for Women to pursue these ideas in a practical form.[40]

While Haslett was enthusiastic about this new popular focus on electricity in domestic life, calculated to create more employment for women, Lady Parsons was not, preferring a focus on more formal engineering.[41] Nevertheless, Lady Parsons agreed reluctantly to host the EAW's first meeting in her London home on 12 November 1924, with Sir Charles Parsons present and Margaret Partridge, among many others, in attendance.[42] There it was agreed that the EAW's main aim would be to use electricity to remove "drudgery" for women at home, with the implication that opportunities for emancipation beyond the home would thereby be opened up.[43]

In the face of Lady Parsons's coolness, Haslett recruited the eminent (and first) female Conservative MP to her cause – Lady Nancy Astor – to become the EAW's first president. In her first address to members in 1924, Lady Astor was amused to see that women were "expecting to be emancipated by electricity." For her, the "most difficult thing in a house was a man," and electrical equipment "would not get rid of that."[44] As we shall see shortly, the behaviour of men was indeed a problem for the EAW. But the biggest challenge initially was how the EAW's domestic focus was seen by Lady Parsons as a distraction from the WES's purposes as the EAW's parent body. Matters came to a head in planning for the first international conference of Women in Science, Industry and Business at Wembley in 1925. While Haslett wanted the WES contribution to be Margaret Partridge's paper on "The Production and Distribution of Electricity," Lady Parsons objected. After a major stand-off at a WES meeting that spring, Haslett won over the membership to her view: unused to such outright defiance from the organization that she had nurtured, Lady Parsons withdrew her WES membership and financial support, and broke links permanently with both the WES and the EAW.[45]

Both organizations survived this traumatic episode, but afterwards Haslett stayed on as the WES secretary only until 1929, owing to the increasing demands of serving the EAW as its first director. Instead she focused on mediating between skeptical female householders and the electrical industry that rather evidently needed to convert women at home into consumers of electrical energy. As the *Woman Engineer* reported on Haslett's own speech to the Wembley conference in the summer of 1925:

> It is quite apparent in all discussions of electrical power
> for domestic purposes that one of the chief difficulties the
> manufacturer has to face, and one which the engineer seems
> incapable of grappling with, is the attitude of the average
> housewife towards these questions. In spite of propaganda and the
> work of clever salesmen there still exists a wide gap between the
> suppliers of "juice," the makers of electrical appliances and the
> woman who is the potential customer. It is impossible to expect
> commercial enterprise alone to fill up this hiatus, and education
> on simple constructive lines is necessary if the woman in the
> home is to realize to the fullest extent the value of electrical
> development for domestic purposes.[46]

Haslett calculated shrewdly that this would be a much more easily accomplished project, and indeed much more attractive to commercial sponsorship by electrical suppliers. Indeed, the broader enthusiasm of industry for the EAW was evident after Haslett's Wembley lecture. Requests swiftly came in for public lectures and visits to power stations from managers of electricity departments, as well as support from the British and Allied Cable Manufacturers Association and from the minister of transport. So rapid was the EAW's growth, with branches soon opening in Glasgow and Manchester, that the EAW moved out of the WES building in 1927 to relocate above an electrical substation of the Kensington and Knightsbridge Electricity Lighting Company.[47] Much of the EAW's story henceforth has been well told by several historians – Peggy Scott, Suzette Worden, and Carroll Pursell – so no further restating of the relevant institutional details of its forty-two year history is needed here.[48] What I shall do instead, however, is explore the quite complex gender dynamics of its operations, as manifested in the *Electrical Handbook for Women* of 1934.

The EAW and the Gendered Management of Electrical Energy Consumption

The EAW's operations have been described by Pursell as "conservative modernism" and by Worden as a form of low-key feminist endeavour. In a sense both are correct given the mixed demographic involved in the

EAW. Yet more importantly, there was a fundamental aspect that both overlook, which was the deep reluctance of many female householders to install *electric* power in their homes – it was, after all, more expensive, less reliable, and less pleasant to work with than gas, coal, or paraffin, assuming even that in interwar Britain a householder was lucky enough to be able to access a supply from a mains supply in a nearby street. So it was by no means an organization engaged in a mundane propaganda activity to increase consumption of electrical energy – both the hearts and minds of female consumers had to be won over. And this required more than just enabling teams of sales staff and demonstrators (all women) of the sort long used in the gas industry, as observed by Anne Clendinning. The development of a whole schema of educational classes and EAW qualifications by examination marked out the domestic consumption of electrical energy as a form of higher scientific culture, run by specially trained female experts – a phenomenon not seen in the nurturing of gas or other utility consumers. It was for such scientized trainers and learners that the *Electrical Handbook for Women* was specifically written.

Such was not evident at all, however, to some of the men who met Haslett on her trips to British towns and cities in setting up EAW branches. Whereas in most locations she was met by all-women delegations of former suffrage campaigners, Women's Institute members, and a range of female professionals, in one city (unnamed) she met a group of men from the electrical industry who had assumed that the EAW was set up only to supply them with female household customers. She advised these men to go away and send their female kin to form the local branch instead.[49] The bilateralism of her strategy is evident in the very different kinds of correspondence that she undertook as EAW director. Haslett's aim to get EAW members so efficient at housework that they could also hold independent careers required the electrical industry to fund the EAW's training schemes. In return for such funding, that (still largely male) industry expected the EAW to bring it more female consumers.

While aiming to please both parties in this transaction, informed observers could readily perceive an ambiguity. Indeed at times some quite radical translations between different agendas were required to bring industry together with consumers.[50] For example, in securing a £50 annual subscription to the EAW from the Incorporated Municipal Electrical Association (IMEA) in April 1928, Haslett had to formally apologize that one EAW lecturer had recently made a public declaration that electric cooking was "a good deal too expensive" to be practicable. Haslett's dexterous handling of this diplomatic mishap involved her explaining that the lecturer had been referring in fact to her own off-grid country house in ways that were allegedly taken out of context by the press. This representation worked, for by 19 May, Haslett had the IMEA's annual instalment of a £50 cheque.[51]

Conversely, Haslett had to deal tactfully with EAW members' responses to electrical supply companies who were trying to change the electrical consumption behaviour of homemakers. This was particularly the case in the company campaign to promote the use of mains-powered dishwashers and vacuum cleaners in the daytime slack hours, when little electricity supply was being drawn upon. From the point of view of suppliers, this was a matter of needing to balance the daily pattern of the consumption "load" upon the electricity supply in a more efficient way. However, Haslett noted that she had to inform supply companies that female energy consumers did not take well to being represented as a "domestic load." On other occasions, when electrical suppliers and allied capitalists were not in earshot, she told such (prospective) consumers that electricity was their route to female emancipation:

I do not think that the women's world has yet realised that the machine has really given women complete emancipation. With the touch of a switch she can have five or six horsepower at her disposal; in an aeroplane she has the same power as a man.[52]

After a decade of building EAW centres around the UK, overseeing the national growth of electrical housecraft courses, and negotiating with the electrical industry for finance, the EAW was by 1933 operating independently of the WES, and so was incorporated as a legal body by the Board of Trade. Its three publicly declared goals were education and training for women in electricity, promoting research on women as consumers of electrical energy, and running formal examinations on knowledge of domestic electricity to provide qualifying credentials for professional female workers.

Following some particularly astute fundraising by Haslet from the electrical industry, by 1934 the EAW was in a new position to promote its goals. The annual report of April that year – the year that marked its first decade – showed a robust activity not only in organized visits to power stations in the UK's metropolitan districts, but also a very successful conference on "Electricity and the Countrywoman" in Birmingham the previous spring, in which prizes were awarded to those members who had introduced dozens of newcomers to the EAW. Many civic branches of the EAW were holding meetings to discuss the design and efficiency of electric cookers, space heaters, laundry equipment, water heaters, and so on, with questionnaires that would send members' opinions directly back to the manufacturers. The Electrical Housecraft School had been the scene of "many varied activities," such as three lecture courses dedicated to "housewives," a theoretical course for teachers, and a new course for "Junior Demonstrators and Saleswomen." Moreover, the industrial Electrical Development Association and the Electrical Lamp Manufacturers Association had sponsored three-day courses for electrical demonstrators, along with numerous sessions for students and teachers visiting from schools across the nation.

The success of the EAW's educational work was demonstrated most notably in the many requests from around the country for syllabi that students could use to prepare for the EAW certificate examination. In some ways most significant for the new range of publications that the EAW was about to launch was a new diploma qualification for "teaching electrical housecraft," to complement the existing syllabus for "demonstrators and saleswomen," and correspondence courses for members who were located remotely.[53] The new publication in development was announced in the annual report as follows:

> "Electical Handbook for Women"
> In preparing this [Correspondence] Course, it was found that no textbook covering the necessary ground was available, and as it was realized that educational authorities in planning Electrical Housecraft Courses, had experienced the same difficulty, the Association decided to compile a handbook which would be suitable for Demonstrators or Teachers, studying the rudiments of electrical technology in connection with their work. With the co-operation of many experts in various branches of the electrical and educational worlds, this handbook has now been prepared and is published by Messrs Hodder & Stoughton under the title *The Electrical Handbook for Women.*[54]

This volume included eighteen chapters, mostly written by the EAW's female technical staff – notably Margaret Partridge, who wrote the opening six chapters on "General Principles of Electricity," with other EAW teaching staff writing the section on "Practical Domestic Applications of Electricity" and the final section on "Teaching and Demonstrating"; male authors were only used for legal and installational matters, and for the less obviously labour-saving wireless radio set. At the end of the work were thirteen pages of examination questions for readers to test their knowledge against the EAW's various formal qualifications in electrical housework.

As Haslett's sister later noted, this *Electrical Handbook* was "an immediate success," with some dedicated EAW members so devoted to

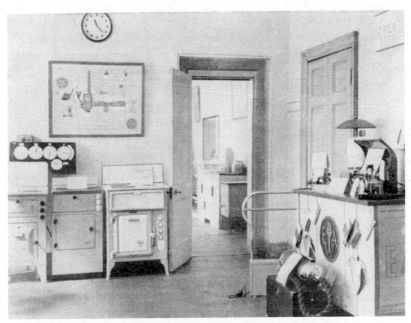

5.2   Frontispiece for the Electrical Association for Women's *Electrical Handbook for Women* (1934), edited by Caroline Haslett

it that they "slept with the book under their pillows"; material from it was extracted for posters and charts to be distributed across the UK's secondary schools.[55] Indeed, such was the popularity of the *Handbook* that Worden notes that in the first year of publication, the EAW sold no fewer than 33,000 copies in its first year – clearly meeting a demand that was at least ten times the membership of the EAW.[56] This demand continued at least until a sixth edition in 1965, the last to be published under the name *Electrical Handbook* for women.[57] By 1986 the EAW was wound down, having only partially completed its work of converting most of Britain's households to electrical consumption: many still preferred to use gas for cooking and heating. As Pursell points out, with so few younger members joining the EAW in the 1980s, the hitherto supportive electrical industry effectively closed it down by refusing to grant further financial support or to recognize EAW qualifications for work in the industry.

Ironically, perhaps, Haslett – in collaboration with EAW lecturer Elsie Edwards – helped to subvert the business of EAW qualifications with their 1939 volume *Teach Yourself Household Electricity*: this work signalled no necessity for any formal EAW training or examination to accredit its readers.[58] Although assembled for use in wartime, when EAW training activities were suspended for other forms of national defence work, the message of this book was clearly that any woman (or man) could find their own way – in wartime or peace – to perfecting the practice of electrical housecraft.

Conclusions

In this chapter I have shown how Haslett and fellow EAW workers mediated the obvious mid-twentieth century gap between electrical energy suppliers' wish to win over prospective consumers, and (presumptively) female householder reluctance to consume energy via electrical supply. Accepting industrial sponsorship to support their promotional activities, the EAW helped to normalize the housecraft of electricity as a highly skilled and prestigious "scientific" activity. Although they did not directly fulfil Haslett's early promises of female emancipation through reduced drudgery, a widespread early disinclination among householders to adopt domestic electricity was significantly overcome by EAW efforts. Indeed, the sheer scale of the EAW labour that was involved in this enterprise brings into question historical accounts that assume that householders necessarily *wanted* electricity to energize their houses as soon as it was available to them. Although the EAW ceased operations in 1986 without ever normalizing the all-electric house – the mixed gas-electric domestic economy instead remains the norm – the EAW did at least achieve Haslett's goal of creating a schema of technological employment for women in the electrical domain at a time when the WES was finding it much more difficult to create careers for women engineers.

Looking beyond the intertwined stories of the WES and the EAW at some broader theoretical questions, how do the stories of women in this chapter enhance our broader historical understanding of energy consumption and energy transitions? In three interrelated ways it has brought some important but hitherto occluded women's roles to the fore, subverting the conventional gendered stories of self-sufficient male dominance in energy supply provision. Specifically, I have shown how women as engineering spouses could be involved in developing new technologies of power supply, how WES activists challenged gender norms of who could practice as an engineer, and how the EAW challenged the gendered norm of male experts as the authorities and teachers on new electrical technology. That lattermost point was critical for the gendered authority relations of domestic electrification: only *female* authority figures could effectively persuade female householders of the merits of becoming electricity consumers.

Notes

1   Moir and Haslett, "Preface," *Electrical Handbook for Women*, 18.
2   On Bip Pares, see Yasmin Namdjou, "Bip Pares – the Forgotten British Art-Deco Icon," Sulis Fine Art, 21 January 2016, https://www.sulisfineart. com/blog/Bip-Pares-Artist-Spotlight, and "Pares, Ethel 'Bip' (1904–1977)," Radnorshire Fine Arts Ltd, https://www.radnorshire-fine-arts.co.uk/brand/ pares-ethel-bip-1904-1977. There is as yet no entry on Bip Pares in the *Oxford Dictionary of National Biography*.
3   For recent discussions on consumer history that does not discuss the role of female agency in developing energy consumption, see Trentmann, ed., *The Making of the Consumer*. For the German context, see Möllers and Zachmann, eds., *Past and Present Energy Societies*.
4   Gooday, *Domesticating Electricity*.
5   Haslett, *Electrical Handbook*, 16.
6   Hannah, *Electricity before Nationalisation*; Hughes, *Networks of Power*.
7   Roberts, "Electrification."
8   Haslett, *Electrical Handbook*, 17.
9   Ibid., 19–21.
10 Pursell, "Domesticating Modernity." For women's resilient *non-consumption* of electric energy, see Sprenger and Webb, "Persuading the Housewife to Use Electricity?"
11 Worden, "Powerful Women."
12 Schwartz Cowan, *More Work for Mother*, 100–1.
13 Hughes, *Networks of Power*. The sole mention of a female practitioner (p. 150), of Hertha Ayrton as an authority on the electric arc light, was apparently the work of Hughes's spouse Agatha Chipley Hughes – see Allen, *Technologies of Power*, 2–3.
14 Sayer, "His Footmarks on Her Shoulders."
15 The obvious exception is Laura Annie Willson, who ran a building company with her spouse in Halifax, West Yorkshire, arranging house construction; she was involved in both militant suffragism and then later was another cofounder of the WES. See Liddington, *Rebel Girls*, although note that Liddington persistently misspells "Willson" as "Wilson" throughout.
16 Canel, Oldenziel, and Zachmann, *Crossing Boundaries, Building Bridges*. Ironically, Schwartz Cowan's preface to this volume does not connect the phenomenon of women as professional participants in engineering to her study of women as marginalized homemakers in *More Work for Mother*.
17 Gooday, *Domesticating Electricity*; Gordon, *Decorative Electricity*.
18 Gordon, *Decorative Electricity*, 159–60.
19 Ibid., 153. The reference to "drops of golden light" might be to a line in Percy Bysshe Shelley's *Prometheus Unbound* (London: C & J Ollier, 1820): "By the

swift Heavens that cannot stay, / It scatters drops of golden light, / Like lines of rain that ne'er unite; / And the gloom divine is all around."

20   Gordon, *Decorative Electricity*, 159.
21   Ibid., 162–5.
22   Ibid., 168.
23   Ibid., 169ff.
24   Ibid.
25   More wide-ranging discussion of this can be found in work by Sophie Forgan and Graeme Gooday, currently under revisions for *Technology & Culture*.
26   Tyne and Wear Archives, Newcastle: "Lady Katherine [*sic*] Parsons" file L/PA/254.
27   Houston, "The Hon. Lady Parsons." I have thus far been unable to find out any details of who Mary Houston was or how she came to have a connections to the North East Coast Institution of Engineers and Shipbuilders. For more discussion of the role of women in the Parsons family engineering business, see Heald, *Magnificent Women*, especially 17–53.
28   Appleyard, *Charles Parsons*, 104.
29   See "Obituary: The Hon. Lady Parsons," *Heaton Works Journal*, December 1933, reproduced on the Women's Engineering Society website, https://www.wes.org.uk/content/obituary-hon-lady-parsons.
30   For Atalanta, see "Atalanta Ltd," *Grace's Guide to British Industrial History*, https://www.gracesguide.co.uk/Atalanta_Ltd.
31   Houston, "The Hon. Lady Parsons."
32   Ibid.
33   Messenger, *The Doors of Opportunity*, 22–7. Messenger notes that the senior suffrage campaigner Emmeline Pankhurst expressly disapproved of Caroline's engineering ambition. This biography by Haslett's sister, Rosalind, is drawn directly from the substantial collection (thirty-two boxes) of Haslett papers in the archives of the Institution of Engineering and Technology, London. A short summary of Messenger's work can be found here: "Caroline Haslett," Engineering Timelines, http://www.engineering-timelines.com/who/haslett_c/haslettCaroline4.asp.
34   See the *Woman Engineer*'s first issue, 1919.
35   Partridge, "The Direct Current Dynamo," 26–7.
36   Haslett, *Electrical Handbook*, 19–21.
37   *Woman Engineer* 1, no. 13 (1922); Locker, "Partridge, Margaret Mary."
38   Messenger, *The Doors of Opportunity*, 30–3. Lady Beilby was presumably the spouse of the industrial chemist Sir George Beilby FRS, who together were well-known for supporting women's entry to the professions. See HCHC, "George Beilby," https://royalsocietypublishing.org/doi/pdf/10.1098/rspa.1925.0144.
39   Messenger, *The Doors of Opportunity*, 37–46.

40 Ibid., 53.

41 Ibid., 51.

42 Initially it was branded as the Women's Electrical Association, but to avoid confusion with the by-then well-established Workers Education Association, the name was soon reordered to yield the abbreviation EAW.

43 Messenger, *The Doors of Opportunity*, 50.

44 Ibid., 38–9.

45 Ibid., 54. This topic is much discussed in the Haslett-Partridge correspondence of March to May 1925. Caroline Haslett-Margaret Partridge correspondence in the Caroline Haslett papers, IET Archives, NAEST 4.91 folder I, covering the year 1925.

46 Messenger, *The Doors of Opportunity*, 48.

47 Ibid., 56–61.

48 See a firsthand account in Scott, *An Electrical Adventure*.

49 Messenger, *The Doors of Opportunity*, 63.

50 Ibid.

51 Caroline Haslett papers, IET archives, NAEST 33/2.5.1-5.

52 Quoted in P. Scott, *An Electrical Adventure*, 1.

53 Haslett, et al., *Ninth Annual Report of the Electrical Association for Women*, Caroline Haslett papers, IET Archives.

54 Ibid., 7. Hodder and Stoughton was then renowned as a publisher of educational volumes, and later of autodidactic volumes.

55 Messenger, *The Doors of Opportunity*, 71–2.

56 Worden, "Powerful Women," 147.

57 Later editions dropped the gender-specificity in the title, and the ninth edition was published as a generic guide to household electricity in 1983.

58 Haslett and Edwards, *Teach Yourself Household Electricity*.

6

Brown Bread and Washing Machines: Nostalgia and Perspective in Irish Women's Experiences of Rural Electrification

Sorcha O'Brien

Introduction

> I consider that we were a very privileged generation because we saw both sides. Like we saw the time when we had very little and we had no facilities or no gadgets.
>
> Noreen Durken, interviewed by Mary O'Reilly, 13 December 2016[1]

This chapter explores a number of interviews completed between 2016 and 2018 with Irish women in their seventies and eighties about their memories of rural electrification in Ireland. Widely available electricity came to rural Ireland between the late 1940s and mid-1960s, at a later date than we have seen elsewhere. Firstly, the chapter will frame the rural electrification project itself in Ireland, as there were specific economic, religious, and political contexts which influenced implementation and uptake. As women's voices have often been not recorded in the official archives that inform the standard histories of energy, it also explores the importance of oral history as a method that both complements and extends more traditional design history approaches. While oral histories allow us to hear the previously excluded voices of ordinary women, it has also been important to explore the role that nostalgia plays in the interviews, and the processes of memory that inform the women's commentaries on different aspects of life both pre and post rural electrification. It considers how Irish women negotiated this energy transition, and explores whether the move to electrically powered housework echoed Ruth Schwartz Cowan's arguments in *More Work for Mother*, that the impact of "labour saving" devices on women's housework was nowhere near the level which was promoted.[2] These interviews highlight a range of issues about memory and progress, in the ways in which Irish rural women negotiated this energy transition, its effect on their daily work, and gender roles.

Rural Electrification in Ireland

The official Rural Electrification Scheme was launched by the Electricity Supply Board (ESB) in November 1946. This was followed by a move from the construction of hydroelectric generation to thermal power stations, typically fuelled by peat, continuing a deliberate post-independence policy of exploiting indigenous energy resources, rather than relying on coal imported from Britain.[3] The Scheme was endorsed by a succession of Fianna Fáil and "rainbow" coalition governments throughout the 1950s and 1960s, with the common aim of improving rural Irish living conditions and reducing poverty.[4] Providing electricity and running water on the farm and in the farmhouse was intended to help reduce emigration, which peaked in the 1950s, with nearly half a million people emigrating out of a population of three million, a large proportion of which were young women.[5]

The Rural Electrification Organisation, or REO division of the ESB, worked through 792 separate areas, signing up customers, installing poles and wires, and then officially "switching on."[6] This started on a small scale in the late 1940s, it ramped up during the 1950s, and the main project continued until 1964, with a post-development plan returning to areas to reconnect stragglers and "uneconomic" areas such as the islands and some mountain valleys continuing until 1973.[7] Most areas were comprised of between 200 and 400 houses, which would have been initially canvassed by ESB area organizers, who were members of the REO engineering staff and exclusively male. This was then supplemented by public lectures, moving from chalkboards and lantern slides to short films purchased from the Electrical Development Association in the UK, as well as working demonstrations of farm and household appliances.[8] The lack of knowledge about electricity amongst the rural population extended to its use and handling, as well as appliances, with numerous stories of people covering the sockets in case the electricity leaked out, and fears about fire in particular being the reason for many "refusals."[9] In addition, the economic situation and uneven farm incomes meant that cost was a central worry for many households, particularly in the context of paying a standing charge on top of existing domestic rates.[10]

6.1   Electricity Supply Board staff demonstrating domestic
appliances, Muintir na Tíre Rural Week, Mullingar, County Offaly,
May 1954

Despite these concerns, the growing consumer base meant that
ownership of electrical appliances rose steadily throughout the two
decades.[11] During this period, political power swung between the the
two main centre-right parties, Fianna Fáil and an interparty "rainbow"
coalition led by Fine Gael, both of whom were still working with
conservative financial models.[12] A small drop in appliance sales between
1956 and 1958 demonstrated the sensitivity of the Irish appliance market
to trade tariffs, as it was caused by the introduction of extra tariffs on
"non-essential consumer goods" by the first interparty government. In
contrast, the policies of the following Fianna Fáil government were
heavily influenced by Seán Lemass, who was initially Tánaiste and
minister for industry and commerce, and later Taoiseach.[13] Lemass took
a very different approach to that of his predecessors, introducing free

trade policies from 1958 onwards. These policies included lifting import duties on electrical appliances, reducing the cost of appliances imported from Britain, Sweden, and the Netherlands.[14] The market penetration for larger appliances remained low at the end of the 1960s, despite the existence of hire purchase schemes with the ESB, local shops, and department stores, although there was also some competition from gas cookers and stove-top kettles, as well as coal or gas ranges. The highest sales were of small appliances such as electric irons and kettles, largely due to their lower cost of £1 or £2, given the poor economic situation of the majority of the Irish population at this time – while the majority of rural customers were being wired for the first time with state-of-the-art wiring, the high up-front costs of £30 to £70 for such appliances was added to concerns about running costs.[15]

Irish Women and the Politics of Electrification

Irish women had played a radical role in the politics of independence during the 1910s and 1920s, with the women's political organization Cumann na mBan including some of the most radical voices in the movement.[16] Despite full female suffrage being achieved upon the creation of the Irish Free State in 1922, the backlash against this radicalism saw a redefinition of women's role in the new state as being wholly located in the domestic sphere.[17] This was enshrined in De Valera's Constitution of 1937, which was hugely influenced by the Catholic bishops, particularly the Archbishop of Dublin John Charles McQuaid. Article 41.2 of the Constitution refers only to women in terms of their roles in the home and as mothers, making the assumption that all women are mothers and that their "duties in the home" take precedence over any other form of labour.[18] This cultural identification of women as mothers was the reason for the 1932 "marriage bar," which excluded married women from public service work, including teaching and nursing, until Ireland's accession into the European Economic Community in 1973.[19] This legal status ensured that the number of working women remained small, restricted to a limited number of professional women and a much larger group of working-class women. It also emphasized the rural housewife, or *bean an tí*, who looked after the household while her husband ran a farm or small business. Gender roles were quite rigid, with women who did not conform to the maternal ideal mostly either emigrating or ending up in mother-and-baby homes or Magdalene laundries. The lack of contraception meant that families were generally large and the main focus for women was housekeeping and child-rearing, with

6.2   Housewife of the Year, Mrs Annie McStay, with her family,
on the cover of *Woman's Way* magazine, November 1969

cooking, cleaning, and washing high on the agenda. A small measure of
economic independence came from traditional home enterprises, such
as knitting or rearing chickens, where clothes, eggs, and meat produced
for the family could also be sold for a small independent cash income.[20]
    This was the context for the Electric Irish Homes research project,
which set out to investigate and document the role that domestic
electrical appliances played in rural Irish women's lives during this
time period.[21] While employing design history methods to analyze
the appliances themselves and the changing kitchens in which they
were located, and more traditional historical research in national and
corporate archives, the project included an oral history component,
gathering audio narratives from women in their seventies and eighties
who remembered rural electrification in Ireland. The interviewers,
including myself, were all Irish or Scottish women in their twenties to

sixties, resident in Ireland, who either already had experience with oral history interviewing or were given training as part of the project, and in most cases were known to the interviewees. The fifty-three female interviewees were partially recruited through the Irish Countrywomen's Association (ICA), a women's organization which was very active in supporting rural electrification in the 1950s and 1960s, and partially through personal contacts of both the participants and interviewers.[22] The recruitment process aimed to gather a spread of interviewees from across the Republic of Ireland, with participants from across all four provinces, and clusters in Mayo, Monaghan, Kilkenny, West Cork, and Galway.[23] It also aimed to explore a broad spectrum of rural Irish society, with interviewees describing homes which ranged from small cottages to Georgian country houses, and encompassing households on both small and large farms, as well as those with husbands employed in village shops and town banks. While the size of the sample meant that a true statistical crosssection of the population was not possible, the aim was to be as representative as possible of the variations of lived experience amongst the demographic of rural women now in their seventies and eighties.

Although they were intended to explore the introduction of rural electrification to the interviewees' homes in the 1950s and 1960s, the interviews often focused on the purchase and use of electrical "labour saving devices," including electric cookers and washing machines. This focus on "labour saving devices" came about because existing research concentrated on the work of organizing and installing the electricity network, and little extant material focused on the impact of rural electrification on the Irish home and domestic life.[24] For example, the ESB Archive oral history collection contains a number of interviews with engineers who worked on the scheme, with four interviews out of twenty-five with women, one of whom worked for the ESB administering the Scheme. The Irish Life and Lore collection and the Digital Repository of Ireland Life Histories and Social Change Collection both hold some references to rural electrification, but only as brief mentions within longer life-history interviews.[25]

The average Electric Irish Homes interview lasted between half an hour and forty-five minutes, directed in a semistructured manner towards the specifics of everyday life before and after electrification, and included the experience of both the women themselves and often their mothers, as well as related topics such as involvement with the ICA, raising chickens, or domestic textile work. This method of research broadened the range of material available on the subject of electrification, but the nature of oral history meant that it also complicated the official narrative, providing a multiplicity of voices and experiences that neither newspapers, governmental, nor corporate sources reflect. As Paul Thompson outlines, these official sources present "the standpoint of authority," whereas oral history can call "witnesses" from the unprivileged and the silenced, such as the women who cooked, cleaned, and raised children in rural Ireland, but whose opinion was rarely consulted by the authorities.[26] However, as Alessandro Portelli cautions, we need to be careful not to assume that these women were purely "speaking for themselves," given that the interviews are shaped by both interviewee and interviewer, as well as the time in which they are recorded, and the knowledge that the interview is part of an academic research project, leading to both a museum exhibition and an academic publication.[27] This subjectivity influences the stories that are told, as well as the way in which they are told, and the awareness of the audience influences the insights and reflections. These Irish women demonstrated very little sense of themselves as historical actors, but did share a sense that old practices and artifacts are important and should be recorded and remembered.[28]

Lovely Bread and Nostalgia

KATHLEEN MCQUILLAN: I well remember my mother had a small little white can and she'd put a pint or two of milk to one side in that little can to curdle and thicken and that made lovely bread. She put it in a round oven, a good big oven near a foot wide with a lid on it and she'd lift the turf coals out with a pair of tongs and put the coals on top of the lid and give it half an hour or maybe a little more and the lovely round cake would tumble out of the oven [*laughter*].
    GERALDINE O'CONNOR: Lovely, homemade bread.
    KM: Homemade bread and it was lovely.[29]

The phrase "lovely bread" occurs repeatedly in several of the interviews.[30] The slow baking of bread on turf fires resulted in a moist bread that is not comparable to anything commercially available in Ireland today.[31] The emphasis on bread was partly to do with the cultural status of bread baking by women as one of the foundational duties of the home, as a central part of the daily ritual, leading to its clear status in the discussion of pre-electric housework.[32] One of the difficulties of the move to oven baking that many women experienced was the need to adapt recipes in terms of both time and temperature control.

The ESB demonstrators ran a very active program to help assuage these difficulties, including sponsoring a National Bread Baking Competition which is still in existence today.[33] However, many women, such as Noreen Brady's mother, often resorted to ad hoc strategies in order to produce bread of the expected quality.

NOREEN BRADY: Yes, but anyway, to overcome her difficulties after, because she couldn't manage this bread in the cooker, she couldn't get it right, she made a fire outside with the, and ...
GERALDINE O'CONNOR: With the turf and the sticks?
NB: ... with the turf and the stuff and hung the oven outside and ...
GO: Went back.
NB: ... made her bread out on the street [*laughter*] that went on for a wee while. But then she devised a plan, I don't know who, it might have been Kevin Rowley, I don't know who it was at the time, but cutting the legs off the pot oven and put it into the oven ...
GO: Yes, ah!
NB: ... and made her bread and that was a great, that was a great ...
GO: That was a success.
NB: ... success, yes.
GO: Right.
NB: Yes, and so that, that was how she overcome the bread.
GO: So, the bread was saved for all the family?
NB: Yes, yes.[34]

In *The Future of Nostalgia*, Russian émigré Svetlana Boym writes about how nostalgia was originally considered a medical condition suffered by displaced people in seventeenth-century Europe, including soldiers fighting wars and students studying abroad, as well as domestic workers forced to move to find work. It was considered to be a *maladie* which was curable by leeches, opium, and optimally a return home. It is no longer considered a physical illness as such, but a psychological phenomenon that conflates place and time, often experienced as a yearning for a lost and unrecoverable childhood.[35] Boym outlines nostalgia's relationship with the modern idea of time, which is unavoidably linear, and sees it as particularly tied into the idea of progress as something unavoidable and determined. She identifies nostalgia as a way to resist this sense of onrushing time, and particularly to mythologize the unrecoverable past. Hence, no bread, not even from SuperValu, will ever be as good as that baked on a childhood open fire.[36] Rather than considering it as a reactionary "sickness" that produced "bogus history," this chapter follows more recent approaches to nostalgia as an emotional state, in which the contemplation of the past opens up emotions that are as much about the present as the past.[37] As Noreen and Geraldine discuss the ways in which Noreen's mother "saved" the family staple, they are not only rehearsing the centrality of women's work to Irish society, but also reaffirming the intergenerational community links between all three women. The historical energy transition is being mythologized and presented for an outside audience: both the academic researcher organizing the project and potential future listeners (and readers). This aligns with Gary Campbell and Laurajane Smith's discussion of "progressive nostalgia" as "a particular and unashamedly overtly emotional way of remembering that actively and self-consciously aims to use the past to contextualise the achievements and gains of present day living and working conditions and to set a politically progressive agenda for the future."[38] This active form of remembering coping strategies required during an energy transition acts not just as a way to solidify social networks and a sense of community, but to emphasize the role which women played in negotiating this transition in domestic life.

Washing Machines and Emotional Reactions

In contrast to this nostalgia for baking bread, there is a shared recognition in many of the interviews of the sheer hard physical labour of running a household without electricity or, in many cases, running water. The interviews include lengthy descriptions of the efforts involved in washing clothes before the installation of electricity – bringing buckets of water from a well or stream, scrubbing and rinsing by hand, often with older daughters involved. The amount of work involved prompted one interviewee, Noreen Durken, to follow her description with this comparison of the conditions:

> NOREEN DURKEN: And there was no iron. In order to iron you had to stick the thing into the fire, which you had to do several times when you were ironing to try to keep it hot. And so there were no facilities and women had no amenities at all, it was all slavery.
> MARY O'REILLY: All slavery!
> ND: At the time, they did it, they didn't know any different, so they were quite happy doing it, as long as they were able to pay their bills, and as long as they were able to pay for their electricity

6.3   Putting washed clothes through a mangle on a washing machine in a rural scullery, 1950s

when they got the bill! But really and truly, it was pure drudgery, women worked very hard.[39]

This description may be exaggerated when compared with the lived experience of chattel slavery, but it emphasizes the sheer unrelenting nature of daily housework in a home without electricity. The financial uncertainty and fear of bills continued after the installation of electricity, creating another bill to add to the others. The quote also included the commonly voiced sentiment that "they were quite happy," which is often added to such discussions by interviewees in order to highlight a "simple, happy" narrative of pre-electric life, despite the fact that early twentieth-century Ireland was economically depressed, wracked with internecine political violence, and socially controlled by the Catholic Church.[40] The descriptions of the introduction of washing machines focus heavily on the reduction (if not abolishment) of this hard physical labour and the reduction of "drudgery," a term that is used continuously in both the interviews and the archival material.[41] The washing machines of the time still involved a lot of manual work, because, while they were powered by electricity, fully automated models were not introduced to Ireland until the very late 1960s. The housewife still had to attach the input hose to the tap and the drain to the sink, or pour water into the machine from pans heated on the cooker. She then had to wring the clothes through a mangle, to squeeze the excess water out of the clothes before drying could commence.

> CIUNAS BUNWORTH: The amount of drudgery that lifted from a housewife was stunning, from a housewife who boiled water and did the washing and may have had a wringer, to suddenly having something which could put your thing … and there was still work in it and you lifted your washing and it came out and it was … 80 per cent of the water was spun out of it, and you hung it out on the line and you got it dry that day.[42]
>
> MAUREEN GAVAN: It left everybody, I think, happy, because it took the drudgery out of all the hard work that people had to do, like. You couldn't believe, I can always remember my mother saying, "This is heaven" she says, "this has to be heaven," she loved it, she loved it.[43]

The overwhelming emotions expressed in the interviews on the arrival of electricity are relief and delight, which makes sense in terms of a respite from this narrative of "drudgery." This is often expressed in religious terms, and likened to "heaven," particularly by the older generations of interviewees' mothers and grandmothers, who may have been born as far back as the nineteenth century.[44] Such religious terminology echoes the devout nature of the Irish countryside after the Catholic devotional revolution of the nineteenth century.[45] While the emotions are radically different from those associated with the "lovely bread," I would suggest that progressive nostalgia is still at play here. Campbell and Smith describe a sense not just that the past was not perfect, but that it was "hard, difficult and inequitable," which ties in with the repeated references to drudgery, which have survived from the mid-century discussion to the twenty-first century interviews.[46] Presenting the move from turf and gas to electricity as a religious rapture emphasizes the dramatic change in everyday experience which this energy transition represented.

Work and Women's Roles

The decision-making process behind the acceptance of electricity into the home seems to have varied according to age, economic status, and education level, but from our interviews, it typically involved both members of a heterosexual couple, albeit with more power in the hands of the husband. The almost universally positive female reaction was not shared by their male counterparts, particularly farmers, who were particularly concerned about the financial cost.[47] The ESB put a great deal of emphasis on the economies and improvements possible both on the farm and in the home, but many farmers remained unconvinced.[48] As a result, the ICA made use of their guild-based organizational structure and centralized training courses for women in An Grianán to spread the word amongst members about the benefits of electricity, hosting large numbers of demonstrations for their members around the country.[49] In addition to these efforts, Mamo McDonald, an ICA member from the 1950s onwards and later president of the organization, recalled the ways in which the recruitment process worked, as well as the uses of social pressure on recalcitrant farmers:

GERALDINE O'CONNOR: And men were keen to get it in to the farmyard, into the byre?

MAMO MCDONALD: Yes, quite happy to get it. It would be very useful in the byre to get a light in, but they'd think about [getting] it into the house, maybe a light in the kitchen, but like, they certainly wouldn't like the idea of electrifying your house and putting in ... they weren't on for that, they were kind of dragged into it and inveigled into it ... Well, you see, by coming up to [An Grianán], you were getting a cross-section of women from all round the country at the adult education college at An Grianán and the women were coming and a lot of them, their first holiday away would be coming up to An Grianán to do a course, you know, and there was no problem about with filling courses, that time.

GO: And the young women?

MM: The women absolutely, they just loved it, coming there, and then they were going home as disciples, because they saw, you know, all this.

GO: ... and were the girls encouraged to hold back from getting married, is there a story to that?

6.4 "Take a Day Off" advertisement for an English Electric Liberator washing machine, *Woman's Way*, 21 July 1967

MM: Yes, they were. Well, that was the procedure that we were telling the younger members, the unmarried ones, not to marry a farmer unless he put in the electricity for you in the house and had water on tap.[50]

This example of gendered networking and soft power worked to subvert the existing power structures within Irish society, where male farmers were slow to see the point of improving women's lives, even those of their wives and daughters. Lacking political or economic power, the women of the ICA continued to work within existing power structures, but to put social pressure on the younger generation of farmers to improve their houses as well as their land. The promise or refusal of marriage was the one point where young women could actually publicly hold power within a relationship, and the use of this negotiation to include electricity and running water in the marriage contract is significant.

GERALDINE O'REILLY: Yes, I think that nowadays that we, you know, when we have so much and so many electrical appliances in the house in comparison with what, like my mother had and it did, it made their lives a lot easier. But I think now I often wonder how we as women at their age now, how we've benefited from all the appliances, I think we've just found something else to fill in that time that the machine saves us.[51]

Here, Geraldine picks up on an issue highlighted by Ruth Schwartz Cowan's study of the introduction of labour-saving devices to rural American households in the early part of the twentieth century. Schwartz Cowan found that the number of hours of housework actually went up after the introduction of these "labour saving devices."[52] These types of quantitative surveys were not carried out in Ireland in the 1950s and 1960s, but when asked about what they did instead of working, each of our interviewees replied that there was more domestic work done instead. There is no mention of relaxing, playing sports, or visiting friends as alternative activities for filling time "freed up" by mechanized housework, as the commercial advertisements of this time would have had you believe. Similar to the situation in the United States, these minutes and hours were welcomed, but were filled with more work; in this case predominantly textile work carried out in a domestic setting, such as knitting or sewing, mostly done for the family, but sometimes as piecework for extra income.[53]

Perspectives

The interviews demonstrate a definite sense of generational difference between those who grew up before rural electrification and those who grew up afterwards, and the interviewees in their seventies particularly have a sense of straddling both sides of a divide – not the more recent digital divide, but an electric one. Boym sees the act of nostalgia as a longing for continuity, "a defense mechanism in a time of accelerated rhythms of life and historical upheaval."[54] The process of carrying out our interviews was instrumental in unearthing not just attitudes and approaches of the past, but of the present day. As such, the effects of rural electrification are considered by older women, many of them now grandmothers, in the social and economic context of twenty-first-century Ireland.

> MARY O'REILLY: Anything else you'd like to add to that now, Maura?
> MAURA MCGUINNESS: I don't think so. I think that if anyone that's under fifty now heard me talking, they'd be saying, "Oh, that one must be a hundred years old."
> MO: That one must be living in the dark ages! [*laughter*]
> MM: Yeah, that those things could never have happened, you know, that they didn't ...
> MO: Well, you think it's better for them ones now, that never went through any hardship and grow up just with the flick of a switch everywhere.
> MM: Well, I suppose it's better for them, Mary, but this is why I suppose they weren't able to cope with the recession as well as people of my age, because we were reared with not having a lot and we appreciated what we had. Whereas the young people are born with it and they find it very hard to accept if ...
> MO: ... if there's a go slow, anywhere.
> MM: ... if there is a go slow for them, yeah.[55]

After Maura ironically positions herself as coming from Ireland's pre-industrial past, she points out that she has seen both sides of electrified and unelectrified rural Ireland. However, she explicitly emphasizes coping skills developed before the introduction of electricity, which could be dusted off and re-employed in hard times, skills which the younger generations are not seen to possess. This presents an interesting perspective on the hard manual work discussed earlier, particularly in terms of its continuing usefulness as a formative life experience. While Maura shares the value of hard work noted by Smith and Campbell's urban working class, it is not translated into future hopes for educational attainment here. However, it is still a progressive type of nostalgia, in that it recognizes the moral worth of such work, and frames it as a starting point for the future.[56] It creates an individual narrative where drudgery was not completely pointless and avoidable, but was instrumental in developing tenacity and resilience in this older generation of women. Thus, this "women's work" paradoxically becomes a valued and valuable asset to those who performed it, while still not being valued by society at large. Given the current issues surrounding climate change and energy, it does not seem preposterous to suggest that these pre-electric ways of living may become valuable again on a practical level, depending on the trajectory of the next few years.

Conclusion

This work with women's memories of rural electrification has allowed an engagement with the specific ways that heterosexual, usually married, women experienced this energy transition within a traditional domestic environment in rural Ireland. It looks at the value of women's labour in the home, at a time when this entire social group was discouraged from working outside the home, and only valued for this specific type of work. Their generally positive reaction to the arrival of electricity in rural Ireland belies a more complex ambiguity behind "the light from heaven" – while they were promised a cleaner, easier domestic life, there is no doubt that Schwartz Cowan's thesis in *More Work for Mother* does apply here. The reduction of drudgery that electrical appliances allowed was balanced by an increase in other sorts of work, albeit of a less physically taxing nature.

It is also important to consider these interviews situated in their twenty-first-century context, considering the role of memory and the different ways of looking at the past have influenced their stories. There is a marked difference in the ways in which different types of women's work were discussed, with the less physically demanding and more sensually pleasing work of baking acting more as a trigger for nostalgia than that of washing clothes. This difference in reaction fades away amid the positive emotions elicited by the thought of electricity in general, which brought about an epoch-making event for both the interviewees and their mothers, and occasionally their grandmothers. Their reactions engage with ideas about progress in an ambivalent way, in which the coming of the modern world to rural Ireland is seen as a positive development, but it is also viewed in a slightly pensive way when considered from the vantage point of the twenty-first century, leaving the traditional lifestyle of the previous century behind. The reality of climate change, while not explicitly mentioned in the interviews, gives this discussion an added context, in which it might not be possible to continue producing electricity at current rates into the future. The assumption of continuing expansion that is intrinsic to modernity may not continue to have such a hold over energy practices, and the reduction of physical drudgery enabled by rural electrification may not continue further into the twenty-first century. It is within the bounds of

possibility that the pre-electric energy practices of running a household
may become pertinent again, although it may or may not be within
the lifetime of the generation that remembers it. The recording and
collection of such memories may at least go some way to recognizing
these practices, ideas, and approaches for the future, and providing a
model for an alternative future "normality." However, it is important
that such practices are not necessarily so tightly bound to gender roles.

The gendered nature of the decision-making processes in rural
electrification meant that rural Irish women were rarely making the
choice to accept or reject electricity supply on their own. However,
rather than being passive or apathetic, they exercised their agency in the
gendered ways available to them, using personal and social pressures to
influence the decisions of their fiancées, husbands, and fathers. The work
of the ICA scaled this effort up on a national level, through collectively
recognizing the ability of rural electricity to improve their members'
quality of life. Although the ICA worked regularly with the ESB, they did
not following the company line, but developed their own set of social
strategies that relied on peer pressure and the heavily networked nature
of the Irish countryside to promote electricity.[57] Although the arrival of
electricity into the Irish countryside did move the character of domestic
work away from relentless "drudgery," it made very little impact on the
heavily gendered practices of everyday life. Irish women achieved partial
inclusion and success as they developed agency and social power in the
campaign for rural electricity, achieving better working conditions for
themselves and demonstrating an ability to organize. In many ways, this
prepared the ground for the arrival of second-wave feminism in Ireland
in the early 1970s, in which the legal and social structures that shaped
women's lives really started to be challenged.[58]

Notes

1   See O'Brien, "Electric Irish Homes."
2   Schwartz Cowan, More Work for Mother.
3   The first hydroelectric power stations built in the Irish Free State were
    hydro-powered – the 1929 Shannon Scheme was followed in the 1940s and
    early 1950s by two hydro stations on the River Lee, three on the River Liffey
    and three on the River Erne in Donegal. The move to fossil fuels started
    with the oil-powered North Wall in Dublin, opening in 1949; the Marina,
    opening in Cork using coal and oil in 1954; followed by the peat-powered
    Lanesboro, opening in Longford in 1958; and Shannonbridge in Offaly, in
    1965. ESB Archives, "Power Station Portfolio."

4   Ferriter, *The Transformation of Modern Ireland*, 481–2.
5   Ibid., 471–3, 478; Central Statistics Office, "That Was Then, This Is Now."
6   Shiel, *The Quiet Revolution*.
7   Electricity Supply Board, "Electricity Supply Board Forty Third Annual Report," 18.
8   Shiel, *The Quiet Revolution*; Manning and McDowell, *Electricity Supply in Ireland*.
9   Nora from West Cork, interview by Brigid O'Brien, 23 December 2017 (see O'Brien, "Electric Irish Homes"); Eileen Lynch, interview by Mary O'Reilly, 16 December 2016 (see O'Brien, "Electric Irish Homes"). Broader fears and anxieties about electricity are addressed in Ruth Sandwell's chapter in this volume, as well as Gooday, *Domesticating Electricity*.
10 Nuala Stack, interview by Eleanor Calnan, 11 April 2018 (see O'Brien, "Electric Irish Homes"); Joe Kearney, "Before the Light," 53–5; Garvin, *Preventing the Future*, 102.
11 Electricity Supply Board, "Electricity Supply Board Forty Third Annual Report," 16.
12 Garvin, *Preventing the Future*.
13 Taoiseach is the Irish term for the prime minister and head of government, and Tánaiste for deputy prime minister. They come from the Irish language words for leader or chieftain, and their heir apparent, based on the ancient Gaelic tradition of electing a second-in-command who would inherit the leadership, from within the eligible male members of the sept, or clan.
14 Daly, *Sixties Ireland*, 22–5; Ferriter, *The Transformation of Modern Ireland*, 545; Ó'Gráda, *A Rocky Road*, 74–5; Shiel, *The Quiet Revolution*.
15 Electricity Supply Board, "Electricity Supply Board Forty Third Annual Report."
16 McCoole, *No Ordinary Women*.
17 MacPherson, "Women, Home and Irish Identity," 6–8, 34, 134–226.
18 This clause is still currently part of the Irish Constitution, although there were discussions about a referendum to either remove it or make it gender neutral in 2018, though this has not yet happened at the time of writing. "Bunreacht Na hÉireann," article 41.2; Gleeson and Logue, "Referendum on 'Sexist' Reference to Women's Place in the Home Postponed."
19 Daly, *Sixties Ireland*, 151–61; Beaumont, "Women, Citizenship and Catholicism in the Irish Free State," 563–85.
20 Connolly, *The Irish Women's Movement*, 239–41; Clear, *Women of the House*, 214.
21 The Electric Irish Homes research project ran from July 2016 to December 2019, and was funded by the Arts and Humanities Research Council of the UK, through Kingston University and in partnership with the National Museum of Ireland. I was the principal investigator, and the main outputs were the *Kitchen Power: Women's Experiences of Rural Electrification*

exhibition, which opened in the National Museum of Ireland–Country
Life in Castlebar, County Mayo, in July 2019, and a monograph, which is
forthcoming at the time of writing.

22   The interviewees also included six retired female ESB demonstrators, and
four men, two of whom had worked for the ESB in the 1960s and two of
whom wanted to speak about their mothers' experiences. This chapter
will focus on the women with nonprofessional experiences of rural
electrification. The method used here of recruiting interview participants
is referred to as "snowball sampling," in which research participants recruit
other participants for the project, making use of their social networks to
access participants who would otherwise be hard to access.

23   The sample was confined to the Republic of Ireland, as Northern Ireland
was not electrified by the ESB, and had a different political and economic
trajectory throughout the 1950s and 1960s. That said, interviewees from
the border town of Clones, County Monaghan, included some who had
been born in Fermanagh in Northern Ireland, but had married and set
up home in Monaghan. The sample also included two interviewees who
had emigrated from Ireland in the 1950s (to England and Australia), but
who had returned to Ireland upon marriage in the 1960s. The clusters of
interviews corresponded with ICA groups who were actively involved with
the interview project.

24   The main academic source covering part of this time period is Caitriona
Clear's study of women's household work in the early twentieth century, as
both Michael Shiel's history of rural electrification and Maurice Manning
and Moore McDowell's corporate history focus on the ESB and the
infrastructural organization of the scheme. See Clear, *Women of the House*;
Shiel, *The Quiet Revolution*; and Manning and McDowell, *Electricity Supply in
Ireland*, 123–42. More recent work includes that of Rhona Richman Keneally,
"Tastes of Home in Mid-Twentieth-Century Ireland," and Ciara Meehan,
"Modern Wife, Modern Life."

25   Anne Joye, interview by Brendan Delany, 21 March 2003 (see O'Brien,
"Electric Irish Homes"); Irish Life and Lore, "Search: ESB"; Digital Repository
of Ireland, "Life Histories and Social Change Collection."

26   P. Thompson, "The Voice of the Past," 36.

27   Portelli, "What Makes Oral History Different," 56–7.

28   This is in contrast to the Black South African women interviewed by Bozzoli,
"Interviewing the Women of Phokeng," 217–19.

29   Kathleen McQuillan, interview by Geraldine O'Connor, 16 February 2017
(see O'Brien, "Electric Irish Homes").

30   The word "lovely bread" appeared in five of the interviews, always in relation
to baking bread on an open fire in an iron pot, in the traditional manner.

31   Doyle, "A Study of Traditional Hearth Furniture and the Hearth."

32 Shiel, *The Quiet Revolution*, 192.

33 The National Bread Baking Competition started in 1952 in conjunction
    with the ica and the British company gec, and is now run by the
    supermarket aldi with the ica. esb Archives, "National Breadbaking
    Championships"; aldi, "The National Brown Bread Baking Competition."

34 Noreen Brady, interview by Geraldine O'Connor, 17 February 2017
    (see O'Brien, "Electric Irish Homes").

35 Boym, *The Future of Nostalgia*, 12.

36 SuperValu is currently the largest supermarket chain in the Republic of
    Ireland, launched in the 1960s in Cork and still in Irish ownership. Brand
    values emphasize locally sourced fresh produce and the supermarket is active
    in rural communities, particularly sponsoring gaa football and the Tidy
    Towns competition. SuperValu, "About Supervalu."

37 Hewison, *The Heritage Industry*, 10, 143–4; Campbell, Smith, and Wetherell,
    "Nostalgia and Heritage," 609–10.

38 Campbell and Smith, "Nostalgia for the Future," 613.

39 Noreen Durken, interview by Mary O'Reilly, 13 December 2016 (see O'Brien,
    "Electric Irish Homes").

40 Garvin, *Preventing the Future*; Ferriter, *The Transformation of Modern Ireland*,
    450–622; T. Brown, *Ireland*; Daly, *Sixties Ireland*.

41 For example, reo *News*, 3; Fitzpatrick, 8–9.

42 Ciunas Bunworth, interview by Sorcha O'Brien, 13 March 2017 (see O'Brien,
    "Electric Irish Homes").

43 Maureen Gavan, interview by Geraldine O'Connor, 22 February 2017
    (see O'Brien, "Electric Irish Homes").

44 Noreen Durken, interview by Mary O'Reilly, 13 December 2016 (see
    O'Brien, "Electric Irish Homes"); Maureen Gavan, interview by Geraldine
    O'Connor, 22 February 2017 (see O'Brien, "Electric Irish Homes"); Josephine
    Helly, interview by Eleanor Calnan, 11 April 2018 (see O'Brien, "Electric
    Irish Homes").

45 Larkin, "The Devotional Revolution in Ireland," 625–52.

46 Campbell and Smith, "Nostalgia for the Future," 613.

47 "Our Consumers," reo *News*, 1950, 1.

48 Electricity Supply Board, "What a Unit Can Do"; "Save Time and Labour."

49 Ciunas Bunworth, interview by Sorcha O'Brien, 13 March 2017 (see O'Brien,
    "Electric Irish Homes"); Delo from Galway, interview by Sorcha O'Brien,
    6 June 2017 (see O'Brien, "Electric Irish Homes"). The ica were avowedly
    "not feminist" during this time period, but engaged in work improving
    women's lives within the system, which comes under the description of
    "maternal feminism."

50 Mamo McDonald, interview by Geraldine O'Connor, 28 February 2017
    (see O'Brien, "Electric Irish Homes").

51 Geraldine O'Reilly, interview by Geraldine O'Connor, 20 February
    2017 (see O'Brien, "Electric Irish Homes").
52 Schwartz Cowan, *More Work for Mother*.
53 Knitting is mentioned in twenty-nine of the interviews, and sewing
    in nineteen. The knitting of Aran jumpers in particular moved from
    a means of clothing families to piecework during the twentieth
    century, building on existing nineteenth-century movements which
    taught Irish women lacemaking as part of philanthropic efforts at
    economic improvement. See Corrigan, *Irish Aran*. Note that during
    the 1950s, television was only available on the eastern seaboard and
    border counties, where BBC Northern Ireland could be picked up,
    and the Irish national broadcaster RTÉ commenced transmission on
    31 December 1961.
54 Boym, *The Future of Nostalgia*, 10.
55 Maura McGuinness, interview by Mary O'Reilly, 12 December 2016
    (see O'Brien, "Electric Irish Homes").
56 Campbell and Smith, "Nostalgia for the Future," 623.
57 These social strategies were complemented by model and
    demonstration kitchens, training courses, and guild-based talks, which
    will be discussed in more detail in my forthcoming monograph
    Kitchen Power.
58 Connolly, *The Irish Women's Movement*; Connolly and O'Toole,
    Documenting Irish Feminism.

7

Complex Agency in the Great Acceleration: Women and Energy in the Ruhr Area after 1945

Petra Dolata

Introduction

When West German photographer Rudolf Holtappel was looking for a fitting motif for his series on people in the coal mining Ruhr region in the early 1960s,[1] he found what he thought was a perfect shot. As he remembered decades later, at the time, women living in this heavily populated and polluted area would insist on wearing white blouses to spite the coal dust and fly ash. Calling it "inner" or inward-looking "protest" against the enormous air pollution in the industrial Ruhr area (*Ruhrgebiet*), he perpetuated the image of these silently protesting women wearing their best and brightest clothes against sooty smoke-stacks via his black-and-white photographs.[2] The contrast of the women dressed in white and their surroundings could not be starker. This kind of juxtaposition was not only limited to the visual images; it also reflected an internal personal conflict highlighting the ambivalent position of women in this industrial region dominated by coal mining and steel production. In fact, the deliberate choice of white clothing in these dirty environs may have operated as a double act of defiance: against the Ruhr region as an "energy sacrifice zone"[3] as well as against the social stigmatization of the area as uncivilized and working class. In the former case, the women's behaviour constituted a form of environmental protest, while in the latter, it provided social commentary. These women were married to coal miners or steelworkers. Their livelihoods depended on wages earned in the industry which created the pollution. Years later, those same women would protest in support of the industry when structural changes in West Germany's energy mix threatened the closure of coal mines starting in the late 1950s.

7.1   Rudolf Holtappel capturing a woman wearing her white blouse in the Ruhr area, 1961

As much as these women aimed to silently protest the coal smoke because they were mothers worried about their children's health, they also had to endorse the industry as an integral part of their identity as coal miners' wives within a coal culture and economy, their daily lives often spatially confined to the so-called *Kolonien*, coal mining housing estates. Even though they were excluded from working in the mines, they played a crucial role in enabling coal miners to sell their labour and were thus "part of the capitalistic enterprise of mining."[4] For example, until 1970, it was not uncommon for wives to clean their husband's work clothes. Miners usually only owned one set and on their one day off, often Sunday, women were responsible for the washing. This was hard, unpaid work in support of their husband's paid jobs.[5] In the early 1970s, it was estimated by the Max Planck Institute for Working Environment and Human Factors that women spent the same number of calories cleaning their husbands' work clothes as their husbands did working below ground,[6] confirming Ruth Schwartz Cowan's observation that "in reality kitchens are as much a locus for industrialized work as factories and coal mines are."[7] The difference was that women were not paid for their housework. In addition, due to shift work, women had to deal with any issue at home and make important household decisions on a daily basis to ensure the welfare of the family and the cleanliness and respectability of the home. The German coal miner of the 1950s, who was often seen as a symbol of West German reconstruction, literally fuelling the country's economic miracle (*Wirtschaftswunder*), could only make his critical contribution to a prosperous West Germany with the help of his partner at home.

Conceptualizing Women and Energy in the Ruhr Area

Women need to be included in any histories of coal mining in the Ruhr region and not only as appendages. They were housewives and mothers, but they were also women and historical actors in their own right. They were not "house slaves" or "house dragons."[8] Their stories are important not only because they are separate from and complementary to men's stories but also because they highlight the ambivalence and ambiguity of human behaviour in energy systems. Too often these women remained hidden from an energy history that only focused on production and paid energy labour, and thus on activities which were much easier to measure.[9] Conceptualizing energy history to encompass much more allows us to address women's behaviour as both championing and criticizing the coal industry and by extension the existing energy system. On the one hand they were involved in energy production, as they formed an integral part of a coal economy that could only function if women stayed at home and took care of the family's and the miner's needs. On the other, they were active as postwar energy consumers. Working at home may have isolated them but the home was not a separate sphere; it was part of both the production of coal and the consumption of energy, much of which was still provided by coal in West Germany in the 1950s. As Schwartz Cowan reminds us, "whether what they have been doing all day is called 'consumption' … it still takes time and energy."[10] Their work was embedded in a much larger energy regime that was also an economic and social system. Mirroring developments in the wider Western world, their daily lives would increasingly rely on high-energy usage. Household and small consumer energy use in West Germany rose from 32.1 million tonnes of hard coal equivalent units (SKE) in 1950 to 52.4 million tonnes in 1960, almost doubling to 100 million tonnes in 1970 and plateauing at 112.1 million tonnes in 1980.[11]

Understanding individual personal lives as complex practices and narratives of producing and consuming not only coal but other energy carriers allows us to situate women's lives in the Ruhr region within a comprehensive energy system. This energy system includes not only economic and technological aspects but also social and cultural features. In the case of the Ruhr, heating and powering households was closely connected to existing regional coal-based energy production systems. If we understand the Ruhr area as part of such a comprehensive energy system, we can study what has been coined *Strukturwandel* (structural change) as constituting an energy transition. The history of the Ruhr area is deeply embedded in an overarching hydrocarbon-based energy system which was undergoing significant changes after 1945 as both energy production and consumption ceased to be dominated by coal. Two major energy-related developments generated these structural changes in the Ruhr region. Oil was replacing coal, especially in heating markets, and overall energy consumption grew precipitously, not least because of increasing car ownership. As a result, coal production in West Germany declined and, in the long run, only survived due to government intervention.[12] In 1950, 73 per cent of West Germany's energy needs were provided by bituminous coal. That share had plummeted to 19 per cent in 1979, while petroleum's share rose from 5 per cent to 51 per cent and natural gas from close to zero to 16 per cent.[13] While coal miners made up 53.4 per cent of the Ruhr area's industrial labour in 1950, their share declined to 38.2 per cent in 1961 and 26 per cent in 1970. The absolute number of people working in coal mining halved between 1950 (414,008) and 1970 (200,142).[14]

Strukturwandel or energy transition in the Ruhr region is a post-1945 phenomenon and constitutes an important regional substory in the era of the so-called Great Acceleration. J.R. McNeill and Peter Engelke use this term to denote the exponential use of fossil fuels worldwide and the surge in global population growth since 1945 which detrimentally affect the earth system. They argue that the beginning of the Anthropocene, a term first proposed by Paul Crutzen and Eugene Stoermer in 2000,[15] should not be dated to the onset of industrialization and the emergence of an energy regime based on hydrocarbons in the late eighteenth century, but rather to the mid-twentieth century, because since then "the human impact on the Earth and the biosphere, measured and judged in several different ways ... has escalated."[16] Their measurements include the "spectacular climb in fossil fuel use since 1950"[17] and the concomitant exponential growth of carbon dioxide emissions, three quarters of which stem from fossil fuel use, as well as an increasing number of motor vehicles and soaring production of plastics, all of which facilitated surging global population, rising numbers of city dwellers, and skyrocketing use of nitrogen, which in turn fed quickly advancing deforestation and significant decline of biodiversity. In their view, it was the Great Acceleration that "jump-started the Anthropocene."[18] If, as they maintain, "energy seems to be at the heart of the new epoch,"[19] it is crucial to understand the role of human behaviour in this critical time period. However, it is not sufficient to limit this role to either the production of fossil fuels or the surging consumption of energy since 1945. As we saw in the introduction, any discussion of the role of human behaviour during the long transition to massive fossil fuel use also needs to address the complex and all-enveloping nature of an energy system based on hydrocarbons which constitutes and is supported by a social and cultural system.[20] It is ultimately through human choices that energy systems have been shaped. Moreover, this connection between individual agency and larger systemic changes includes individual behaviour that is ambiguous and adapts or rejects existing energy systems.

For a long time, women remained hidden from histories of the Ruhr area. Traditional labour histories of the region tended to focus on coal miners and steel workers and their political movements.[21] Specialized histories of the *Strukturwandel* centred on the deindustrialization of the region and the closing of steel plants and coal mines since the late 1950s without linking these transformations to larger changes in the energy system and disregarding the complex interrelationships of various energy carriers. These two historical perspectives made it easier to hide women's experiences. Women were not seen as playing an important role in either the miner's labour movement or in the politics of deindustrialization. Examining *Strukturwandel* as an energy transition allows us to reinsert them into the region's post-1945 history. It also allows us to examine energy transitions as lived experience. Oftentimes, historians study energy transitions on a much larger level, and spanning decades if not centuries. While considered a minor shift within the so-called mineral energy regime,[22] the change from coal to oil in West Germany was significant and had major repercussions in the Ruhr area. Because people witnessed these changes during their lifetime, they generated their own individual recollections and created personal and local stories.

Women in the Ruhr area were (re)discovered through oral histories and when everyday history became increasingly popular in Germany in the 1980s. Lutz Niethammer pioneered the study of Life Histories and Social Culture in the Ruhr Area 1930–1960 (*Lebensgeschichte und Sozialkultur im Ruhrgebiet 1930–1960*, LUSIR), collecting and analyzing 350 oral history interviews.[23] More recently, Stefan Moitra from the German Mining Museum in Bochum led a research project entitled "Digital Memory Repository: Humans in Coal Mining" (*Digitaler Gedächtnisspeicher – Menschen im Bergbau*), which conducted eighty-four oral history interviews between 2015 and 2018, when the last hard coal mine closed in Germany.[24] Engaging directly with people living in the Ruhr area, the interactive digital project *"Zeit-Räume Ruhr"* (Ruhr Spaces of Time), which was created and curated by the Institute for Social Movements at the Ruhr-Universität Bochum and the Ruhr Museum in 2017 and 2018, featured places of memories in the region along with everyday stories and memories.[25] Since the late 1970s, grass-roots historical workshops (*Geschichtswerkstatt*) have brought people of the region together to discuss their stories, and, influenced by larger feminist movements in West Germany, some of these were specifically dedicated to women's experiences.[26] In the 1980s, Jutta de Jong collected women's stories in the coal mining town of Herten as part of the larger research project "History from Below," led by the German Trade Union Confederation (DGB).[27] Even though these women may not have wielded any formal power and were thus not seen as influential actors in the conventional political history of the Ruhr region, it was felt that their stories would tell us a great deal about how these gendered historical contexts were negotiated on a daily basis. As such, everyday histories of women were not just additions to get a more complete history; they also helped to "decenter conventional historical perspectives"[28] and served as alternative histories.[29]

As Alf Lüdtke, Hans Medick, and Dorothee Wierling have convincingly argued, everyday history does not prioritize individual experience over larger historical developments, but locates historical agency at the intersection of micro and macro levels.[30] Combining autobiographical and personal stories with historical contexts allows us to comprehend how historical structures and individual agency are mutually constitutive. While individual behaviour is determined by existing and durable historical systems, including energy systems, it is through everyday acts where these meanings are reinforced, but also rejected, which may bring about change in the long run. Change can only be understood by looking at all levels of human behaviour and through acknowledging the connections between larger historical shifts and everyday actions. Writing a history from below also means discussing human agency not only as an aspect of historical structures but also as a way of empowering subjects. Examining women as important historical actors will counter the marginalization of female agency in energy history and insert such agency into accounts of the Great Acceleration. What is needed is a more balanced and nuanced approach toward evaluating the roles of women while appreciating that there is no monolithic female experience.

Women in the Ruhr Area, 1950s to 1990s

The Ruhr area in western Germany was dominated by energy production from the second half of the nineteenth century until the 1990s. Located between the river Ruhr in the south, the river Lippe in the north, and the Rhine to the west, it has been one of the most industrialized regions worldwide in the twentieth century. In the immediate decades following the end of the Second World War, this conurbation was the centre of integrated coal and steel production. It was "the heartland of heavy industry in West Germany ... [and] 'the' proletarian region of the Federal Republic," with almost two thirds of its workforce employed as workers and close to half a million miners working in the region's 136 mines in the late 1950s.[31] Coal mining dominated the region's economy and labour market for decades. First misinterpreted as cyclical fluctuations, structural changes in global energy markets – that is, the move away from coal to oil – led the region to slowly deindustrialize and allowed economic diversification to set in from the late 1950s onwards. With regards to women, the *Strukturwandel* included "developmental trends [such as] the increase of female employment, a drop in the marriage rate, decreasing numbers of children, and an increase in the divorce rate."[32] While the Ruhr area was characterized by its working-class population, in the 1950s women often stayed at home. This was in contrast to earlier decades, especially before and after the First World War, when miners' wives had to engage in economically productive work outside the home as domestic workers, farmhands, or shop helpers to make ends meet.[33] Historians of the region have pointed out that the integrated coal and steel industry reinforced "structurally conservative working-class families" which were hostile towards female employment, owing in part to post-1945 West German family policies which favoured a stay-at-home mother and wife.[34]

Women were forbidden to work in the Ruhr coal mines. They did not get dirty. Instead, they were responsible for keeping the rented house (or flat) clean. Not much had changed since the late nineteenth century when in European coal regions "[i]t was women – as housewives or domestic servants – rather than men ... who had to cope with the Sisyphean task of keeping homes clean in smoky industrial towns."[35] Until higher smokestacks dispersed the fine coal dust all over Germany in the 1970s,[36] thousands of housewives in the Ruhr area had to fight the daily battle against the fine black carbon dust that would be brought in by their husbands and children, or that would settle on washings hanging outside to dry, and on windows, sills, and stairs. Every morning they would sweep the windowsills and stairs, and clean the entrance. Hans Dieter Baroth recalls that "clean floors were treated as the housewife's calling card. Saying that one could eat off the floor constituted the highest achievable praise."[37] "The soot not only stuck like flakes but like grease" to every surface,[38] especially on days when the coal plant filters were exchanged. Unlike anywhere else in West Germany, windows were cleaned every week. While the dark frames hid the grime, the glass did not, and "when you had wiped them clean with a cloth, you had to throw it away afterwards. The windows were properly washed with soap suds."[39]

Weekly rituals included the arduous scrubbing and polishing of the floors as well as cleaning the white starched drapes, which were so neatly pleated and which became a centrepiece of resistance against the most visible impact of an energy production system centred around coal. This cleaning work echoed earlier working-class desires for respectability.[40] The obsession with *"strahlend weiß"* (spotless white) was facilitated even further by advertisements of the time. Companies selling washing detergents such as Persil would use this slogan for marketing purposes, and "Persil weiß" ("Persil white") quickly entered everyday language, becoming synonymous with spotless white. Media analyses of the 1950s and 1960s show how "white" became an important trope in postwar West Germany. Everything had to be clean and white. In German, "having a white vest" means, figuratively, having a clean record, and a *Persilschein* (Persil ticket) attested to the clean political past of its holder. After the Nazi period and the Second World War, this is what Germans were striving for, especially in the 1950s and early 1960s under Conservative governments.[41] In the Ruhr area it was even more difficult to keep up the spotless white image of a conservative West Germany in the 1950s. Mining women took pride in the cleanliness of their homes and in being a housewife. The so-called *Kittelschürze* (apron) was their daily companion, representing their self-imposed rigorous cleaning standards. Their cleaning regime was also an attempt to attain respectability.

The nature of coal mining structured the daily and weekly routines at home. Miners' shift work determined the time of meals. Schedules at home were arranged to accommodate the breadwinner. Bärbel Bergerhoff-Wodopia remembers having to wait for her father to come home from the early shift at 3 p.m. before lunch was served. In West Germany at the time, the main warm meal was lunch, and children would normally finish school at around 1 p.m., but in coal miner's homes the rest of the family had to wait for the husband and father to come back from the mine.[42] When miners were working underground, their wives had to make important decisions during the day, including managing finances and dealing with municipal or insurance authorities. Thus, the women's role was not confined to the home but extended beyond it into the public realm, facilitating financial and social power.[43] "Their careful maintenance of their husband's work clothing, boots and even bodies, their management of household resources ... were crucial to the well-being of their families," Valerie Hall argues, examining similar experiences in England, in which "[w]omen made the home a centre of power for themselves."[44] To use Schwartz Cowan's terminology, women's "household labour" supported their husband's "market labour."[45]

As mentioned above, the cleaning of their husband's pit clothes constituted arduous work. Until 1970, when a collective bargaining agreement stipulated that the coal mining company had to pay for, wash, and maintain pit clothes, men would bring these home on their day off.[46] They would clean themselves in pit-head baths, to try to get the worst grime out of their clothes, but it was their wives who cleaned them. One woman, Frau M., remembers, "the clothes, they were black, like the coal! ... There was proper coal dust in it ... you had such a dust cloud, in all the underwear, the suit, the socks ... I can tell you that was quite some work!"[47] Getting rid of the grime and dust was no easy task before the arrival of fully mechanized and powerful washing machines. Not only was it very time consuming, it was also physically exerting, as women carried the heavy soaked clothes as well as large pots filled with hot and cold water, rubbed and cranked the clothes, and hung them up. As another woman, Frau W., recalls, "just imagine, how much energy the woman had to muster!"[48] Often women saved their children's bathwater on Saturday for soaking the pit clothes, and then next morning fresh soap suds were used before hand-washing and cranking them. After washing, they often had to mend the pit clothes in the evening, making Sunday one of their busier workdays. Things became easier as more and more households owned electric washing machines in the 1960s. However, the grimy and oily clothes had to be precleaned and soaked, and often women had to clean the washing machines afterwards to remove coal dust that could damage the washing machine.[49] Another problem was that their husbands still sweated coal dust, so that shirts, pillowcases, and anything that came in contact with the miners would not be properly clean but attained what was referred to as a *Grauschleier*, a grey veil: "That sweat, all the coal dust was still in it! Also, the pillowcases, they always quickly turned black again. They did not really get it out, the men. It remained in their skin, in their pores and in their hair."[50]

As the example of the washing machine shows, women's "household labour" depended just as much "upon nonhuman energy sources" as the "market labour" of their husbands. While housework was spatially isolated work, it was integrated into a larger energy system, creating "dependency on a network of social and economic institutions."[51] However, in the case of the women in the Ruhr, there was an even more direct link between the two forms of labour, as the nonhuman energy source was often the product of the paid work of miners consumed by the women's work at home. Women's use of energy and their decisions at home affected energy production. After 1945, domestic work increasingly

7.2   Women hauling *Deputatkohle* into the cellar, no date

relied on technology that decreased women's manual labour. Often, these new appliances were powered by electricity. Electricity in the region was still generated through the many coal-fired power plants, which were located in the immediate vicinity of the *Kolonien* and visible to all through their air pollution. Thus, while the source of energy may no longer be visible in the home, women consumers in the Ruhr area did not really become distanced from the source of energy, which was, after all, mined by their husbands and then converted into electricity and later district heating in the coal-fired power and heating plants that belonged to the integrated coal and steel companies.

As these consumption behaviours changed, they affected the use of coal and accelerated the shift toward petroleum-based products. In West Germany, household heating and transportation were increasingly dominated by petroleum products. Domestic coal only survived due to industrial energy consumption and government support of coal power generation.[52] Other petroleum products such as plastics substituted coal-based commodities. For example, many of the coal mining houses had so-called *Stragula* floors. These cheap alternatives to linoleum were partly made out of coal, but with the introduction of affordable PVC flooring in the 1970s, *Stragula* was quickly replaced, not least because the new flooring was so much easier to clean: "The *Stragula* was first mopped and then polished. Properly on the knees, floor polish out of the tin right beside you and then the stuff was evenly applied with a cloth."[53]

In addition to flooring – and much more importantly – coal products also provided heating and cooking. In the Ruhr area, part of the miner's pay was the free delivery of coal briquettes for the miner's personal use (*Deputatkohle*). Initially, this led to the delay in modernizing heating in homes in the region. The *Deputatkohle* was delivered to the front of the coal mining houses, creating yet another source of coal dust and dirt that could be carried into the house. It was especially inconvenient to haul the coal into the cellar. Oftentimes children would earn some extra money "shovelling [the coal] into the cellar hole with buckets,"[54] but women also helped, adding to their already heavy workload at home. These coal briquettes also became a form of payment. Many miners did not use their coal, or all of it, and sold it illegally to those households which did, or to coal wholesale buyers who would drive through the *Kolonie*. One woman recalls her husband buying a used kitchen and a children's bed for a tonne of coal each in the 1950s.[55] Thus, one energy carrier, coal, may have been used as payment to modernize the kitchen and buy an electric range, which ran on electricity potentially generated through another energy carrier. In 1950, 52 per cent of electricity in West Germany was generated by bituminous coal. By 1979, its share had dropped to 29 per cent. Oil and natural gas only provided 0.2 per cent of electricity in 1950. Twenty-nine years later, their share had risen to 23 per cent. Consumer choices such as these helped replace coal in the long run.

As well, the so-called economic miracle (*Wirtschaftswunder*) in 1950s West Germany promoted consumption to a heretofore unseen level. Already in the 1950s coal-powered ovens and heaters were increasingly replaced by oil ones, as they were easier and cheaper to operate.[56] Using coal for cooking and heating also meant more labour was needed to provide cleanliness.[57] Even Chancellor Konrad Adenauer had made the switch and explained his preference for oil heaters by complaining that "it was impossible to get someone who will look after my coal heater and take care of getting rid of the slag,"[58] and some cities in the Ruhr area changed to oil-based heating in their public buildings.[59] Not surprisingly, women in the region, who made many of the consumer decisions for the household, followed the chancellor's example. Coal-burning stoves were replaced by electrical ones, "making everything nicely modern all of a sudden."[60] Women also started working part time to be able to buy nice things, such as new furniture and electrical appliances.[61] As the Ruhr area diversified, it attracted more service industries as well as textile and electronics plants, in which they found jobs earning money that could be spent on consumer goods.[62]

Women's consumer behaviour contributed to the stagnation of private household demand for coal. Realizing the loss of household customers, the Ruhrkohle coal sales company started targeting housewives in their marketing efforts. In the early 1970s, they introduced a campaign around two cartoon characters called *Kommissar Gluto* (Inspector Ember) and *Unbehaglichkeit* (Uncomfortableness), whom the inspector is fighting, to advertise the use of coal-burning stoves and Ruhrkohle coal. Using the slogan "Ruhrkohle. Denn Heizwert ist mehr wert" (Ruhrkohle coal. Because heating value brings added value), they targeted homes in West Germany to switch back to coal.[63] Running from 1972 to 1974, the commercial dovetailed with a temporary renaissance of coal as a result of the 1973/74 energy price crisis. Photographs used for advertising suggested a clean and modern form of energy. Coal-burning stoves were presented as economical and easy to handle. The Ruhrkohle sales company promised that they would provide a constant warmth and coziness in the home.[64] While it may not have been easy to convince the modern housewife of the convenience of coal-fired equipment, the introduction of district heating helped to reintroduce coal into their homes indirectly, as most of the district heating was provided through coal-fired heating plants.[65] Still, after 1945, consumers increasingly preferred petroleum for heating because its production, transportation, and storing was much cheaper and provided more energy, while it created comparatively less air pollution.[66]

While environmental concerns became more important in the 1970s, air pollution issues were on the political agenda in the region much earlier. The federal land in which the Ruhr area is located had already introduced anti-air pollution legislation in 1962, signalling that coal production might no longer be prioritized over the environment at any cost.[67] Still, some abatement measures were very slow to be legislated and implemented. According to Franz-Josef Brüggemeier, in the early twentieth century the region was an "area where industry was consciously protected at nature's *expense*" and "rather than reducing pollution, society [had] adapted to the new industrial conditions."[68] This kind of thinking survived into the 1950s, even though criticism of air pollution began to be featured in news outlets of the region and readers sent in letters to demand that municipalities deal with the pollution. In the 1950s, 300,000 tonnes of dust per year were coming down in the Ruhr area. In some very heavily affected areas, it was more than 5 kilogram per 100 square metres within a month. An often quoted study published in 1959 found that children in the region were on average smaller and weighed less than their peers elsewhere in West Germany.[69] As municipalities were discussing how to deal with this health and environmental problem, citizens actively engaged in what were highly public discussions. In letters to the newspapers that were covering a possible legal case against mining companies which were polluting the air, it was especially women who were reminding readers of how it affected their daily lives and especially the health of their children. One woman, Berta S., invited those who were arguing that jobs trumped any stricter pollution laws to live where she was living and "work themselves to death day in day out cleaning away all the dirt. I only clean. And when I think I'm done, everything is already covered in dust again." She also mentioned that she would leave the Ruhr area if she could, but her husband was not able to find work elsewhere.[70] This statement is an important reference to the fact that jobs in the coal industry were still at the top of the wage ladder and it was not easy to get comparably well-paid jobs elsewhere. Another woman, Mathilde E., insisted that "as a mother" she "could only wholeheartedly agree" with newspaper reports to sue companies over air pollution. She had lived in Duisburg for a year and "witnesses with increasing concern how my children are growing up in this poisonous air. My youngest, who has a weak lung, already had to stay in bed a couple of times due to breathing problems."[71]

Even though women protested against daily pollution by cleaning and wearing their white blouses and by submitting letters to the editors, they did not appreciate outsiders portraying their region as dirty. For some of them, the constant cleaning was an attempt to "whitewash" the dirty image of the region that existed elsewhere in Germany. When writer Heinrich Böll, who lived in Cologne just outside the Ruhr area, published his collaborative work with the photographer Carl-Heinz Hargesheimer in 1958, entitled *Im Ruhrgebiet (In the Ruhr area)*,[72] many criticized the fact that the region was not portrayed as being more modern. They were offended by the photos and Böll's description of the region where the "air tasted bitter and the houses were darker."[73] But by 1961 most miners and their wives welcomed the election campaign by Social Democrat leader Willy Brandt who introduced the election slogan "blue skies in the Ruhr area." This was partly achieved through higher smokestacks in the 1970s. By then, a large share of air pollution had shifted to private households, and consumption away from industrial activity in the region. The Ruhr area had changed from being dominated by energy production to energy consumption. Increasing mobility through car ownership and availability of new jobs in the textile and electronics factories or in the expanding service industry contributed to more consumption and pollution.

In the Ruhr area, an energy system based on coal provided men and women with their regional and working-class identities. So, when coal culture was increasingly under attack, women protested and developed a political consciousness. Already in the 1960s, changing energy systems necessitated responses, adaptations, and resilience by those caught up in the structural changes. Changes in West Germany's national energy system led to specific regional policies affecting the production of energy. Coal was being substituted by oil since the late 1950s. This led to high unemployment rates and rapid deindustrialization in coal-producing areas. This *Strukturwandel* led to an early politicization of energy policies in the 1960s and defined domestic energy discussions in the 1970s.

Beginning in 1958, the Ruhr coal mines encountered a sales crisis which was caused by overproduction, replacement by (heating) oil, and competitive imports of US coal.[74] Mines were closed and thousands of miners laid off. More than 200,000 lost their jobs between 1958 and 1964 alone, during a period of overall economic growth in West Germany. Used to being ranked among the top earners in German industries and proud of their role in fuelling the impressive postwar economic recovery and growth, miners held vigils at the entrances to the coal mines and protested, often silently waving black flags. Most famous was the March on Bonn in September 1959, when 50,000 miners protested in the capital of West Germany. However, due to logistical limitations, only a fifth of the 200,000 miners who wanted to join the protest made it to Bonn.[75]

While these protests were mainly carried out by men, women supported their husbands and sons, and from the mid-1960s onwards were more often part of the public demonstrations against the closing of mines. From Gelsenkirchen-Erle in 1966 to Bergkamen and Dortmund-Huckarde in 1967, women would march side-by-side with their men.[76] Once laid off, miners would be accompanied by their wives to the employment centre. All of a sudden, women who spent a lot of time alone at home taking care of the children and the household found that they had to share this authority with their stay-at-home, unemployed husbands. Reinhold Adam recalls how women were saying half-jokingly, "If my husband stays at home, I am getting a divorce." According to him the problem was that these men had worked for thirty-five or forty years, and now "he was staying at home and wants to rationalize everything. He wanted to order his wife how thick the cheese had to be cut, what to shop at the supermarket. And before she had been the boss at home."[77]

Women's struggle for the survival of coal mining in the region continued well into the 1990s. In 1988, the women's initiative Sophia Jacoba was founded to protest the closing of the coal mine with the same name, in Hückelhoven. Women of all ages participated in sit-in protests in front of the Economics Ministry and the Chancellery in Bonn, where they had chained themselves to each other before being forcefully unchained and carried away by the police. Anxious about their husbands' unemployment and paying off their recently purchased homes, the women were supporting the miners' pickets and protest sit-ins. Despite initial successes, Sophia Jacoba was finally closed in 1997. The women's initiative itself was active from 1988 to 1992, and one of its founders, Jutta Schwinkendorf, went into municipal politics, confirming that these protests gave women "training in leadership and created confidence."[78]

Conclusion

This chapter focused on the complex agency of women in the Ruhr area to show the ways in which energy micro-histories from below help us understand the ambivalent behaviours of individual historic actors who navigate their daily lives as both consumers and producers of energy in high-energy societies. The women in the Ruhr area who were living through the changes that would transform the existing energy system were not always aware of the structural nature of these changes. Nor were they aware of their own inconsistent behaviour in simultaneously supporting and challenging the existing coal regime. Women in the Ruhr area formed an integral part of the coal production system through their unpaid work at home, enabling their mining husbands to perform their paid work. On the one hand, these women were protesting the air pollution that was directly linked to their mining husbands' energy production activities, potentially undermining the viability of the industry and thus endangering their husbands' jobs. On the other hand, they were protesting the closure of mines in the late 1950s and 1960s, and then again in the 1980s. They were identifying themselves with a culture tied to coal mining but also embraced the convenience of non-coal-based consumer goods. Their changing consumer behaviour accelerated the downfall of coal as an energy carrier, a downfall that they protested together with their husbands as the changing energy system or *Strukturwandel* led to an increasing politicization of energy questions in West Germany. Women were both perpetrators and victims of the Great Acceleration. They were not actively producing coal; nonetheless, they participated in both sustaining and resisting an energy system based on coal.

Notes

All translations of German sources are by the author.

1   Coal is used throughout the chapter to denote bituminous coal (classification by the United Nations Economic Commission for Europe). In German it is called *Steinkohle* (hard coal excluding anthracite). I am using hard and bituminous coal interchangeably. Another major energy source in Germany, especially for electricity generation, is lignite coal, which is not mined in the Ruhr area.

2   Schneider, "Bild – Geschichte," 12.

3   The concept of the energy sacrifice zone has mainly been used to describe the environmental impact of large-scale coal production in the Appalachian region. See, for example, Jones, "A Landscape of Energy Abundance," 472; Bell and Braun, "Coal, Identity, and the Gendering of Environmental Justice Activism in Central Appalachia."

4   V. Hall, *Women at Work*, 21.

5   The duality of cleaning miners' work clothes as both housework and market labour is also mentioned by Dorothee Wierling in her contribution on everyday stories and gender history. Wierling, "Alltagsgeschichte und Geschichte der Geschlechterbeziehungen," 181.

6   *Einheit*, 7.

7   Schwartz Cowan, *More Work for Mother*, 4.

8   De Jong, "'Sklavin' oder 'Hausdrache'?," 45–50.

9   Schwartz Cowan, *More Work for Mother*, 210.

10 Ibid., 6–7, 193. See also Wierling, "Alltagsgeschichte und Geschichte der Geschlechterbeziehungen," 182.

11 AGEB, "Struktur des Energieverbrauchs." These energy statistics use SKE (hard coal equivalent units). One tonne SKE equals 0.7 tonne of oil equivalent (TOE) or 5 barrels of oil equivalent (BOE).

12 Abelshauser, *Der Ruhrkohlenbergbau seit 1945*; Nonn, *Die Ruhrbergbaukrise*.

13 AGEB, "Primärenergieverbrauch."

14 Petzina, "Wirtschaft und Arbeit im Ruhrgebiet," 507.

15 Crutzen and Stoermer, "The 'Anthropocene.'"

16 McNeill and Engelke, *The Great Acceleration*, 4.

17 Ibid., 27.

18 Ibid., 6.

19 Ibid., 40.

20 This has been described as Petrocultures. However, as such conceptualization focuses on petroleum as the dominant energy source, it is not sufficient to address the significance of other hydrocarbons such as coal, which is still dominating social and economic life in the region and at the time of this case study. Petrocultures Research Group, *After Oil*.

21 Kaschuba, "Volkskultur und Arbeiterkultur als symbolische Ordnungen,"
    199. Jane Long argues that women have suffered from the "double silence
    of both their gender and their class position." Long, *Conversations in Cold
    Rooms*, 4, quoted in V. Hall, *Women at Work*, 6. See also V. Hall, *Women at
    Work*, 19, 173.

22 E.A. Wrigley uses the concepts of organic and mineral economies to
    differentiate between societies characterized by biomass as opposed to those
    based on fossil energy resources. Wrigley, *Continuity, Chance and Change*.

23 Results were published in four edited volumes: Niethammer, ed., *"Die Jahre
    weiß man nicht, wo man die heute hinsetzen soll"*; Niethammer, ed., *"Hinterher
    merkt man, daß es richtig war, daß es schiefgegangen ist"*; Niethammer and von
    Plato, eds., *"Wir kriegen jetzt andere Zeiten"*; and von Plato, ed., *"Der Verlierer
    geht nicht leer aus."*

24 The project was carried out by the Stiftung Geschichte des Ruhrgebiets in
    cooperation with the Deutsches Bergbau-Museum Bochum. It was funded
    by the RAG Aktiengesellschaft, https://menschen-im-bergbau.de/uber-das-
    projekt (accessed 24 November 2019).

25 The award-winning project was commissioned by the Regionalverband Ruhr
    and the Land North Rhine–Westphalia, http://www.zeit-raeume.ruhr (last
    accessed 8 July 2020).

26 Fletcher, "History from Below Comes to Germany"; Lindenberger, et al.,
    "Radical Plurality"; Schöttler, "Die Geschichtswerkstatt e.V."

27 de Jong, ed., *Kinder, Küche, Kohle – und viel mehr!*

28 Lüdtke, "Einleitung," 15.

29 Medick, "'Missionare im Ruderboot'?," 56.

30 Ibid.; Lüdtke, "Einleitung"; Wierling, "Alltagsgeschichte und Geschichte
    der Geschlechterbeziehungen."

31 Goch, "Betterment without Airs," 89–90, 94.

32 Ibid., 94. See also Goch, *Eine Region im Kampf mit dem Strukturwandel.*

33 Günter, *Mündlicher Geschichtsschreibung*, 13, 28–9.

34 Tenfelde, "Vom Ende und Anfang sozialer Ungleichheit," 281.

35 Mosley, *The Chimney of the World*, 55.

36 Brüggemeier and Rommelspacher, *Blauer Himmel über der Ruhr*, 66–72.

37 Baroth, *Aber es waren schöne Zeiten*, 25–6.

38 Frau W., Frau S., Frau J., and Frau M., "'Wir hatten immer den Ruß auf dem
    Fensterbrett' – Putzen neben dem Pütt," dokument 229, in de Jong, *Kinder,
    Küche, Kohle*, 64.

39 Ibid.

40 Skeggs, *Formations of Class and Gender.*

41 Aufenanger, "Der Familienskandal."

42 Bergerhoff-Wodopia, "'Auch die Nachbarn waren alle Bergleute':
    Kindheitserinnerungen an ein Leben mit dem Bergbau." See Stiftung

Geschichte des Ruhrgebiets, "Menschen im Bergbau." Accessed 30 November 2019. https://menschen-im-bergbau.de/menschen/b.

43 de Jong, "Bergarbeiterfrauen," 74.

44 V. Hall, *Women at Work*, 7, 65.

45 Schwartz Cowan, *More Work for Mother*, 6.

46 De Jong, ed., "…*und die Wäsche, die war schwarz, ja, wie die Kohle!*," 4, 9.

47 Quoted in de Jong, ed., "…*und die Wäsche*," 4.

48 Quoted in de Jong, ed., "…*und die Wäsche*," 12.

49 Frau W., Frau J., Frau N., Frau M., and Frau K., quoted in de Jong, ed., "…*und die Wäsche*," 18.

50 Frau W., Frau N., and Frau K., quoted in de Jong, ed., "…*und die Wäsche*," 7–8, 11–12.

51 Schwartz Cowan, *More Work for Mother*, 6.

52 Abelshauser, *Der Ruhrkohlenbergbau seit 1945*.

53 Frau M., Frau J., Frau S., and Frau W., "'Das Bohnern, das war Samstagarbeit!' – Auf Knien über Steinholz, Stragula und PVC," dokument 230, in de Jong, *Kinder, Küche, Kohle*, 66.

54 Schuchmann, "Kindheit in einer Bergmannsfamilie in den 60er Jahren."

55 Housewife, born 1928, "'Doch schon recht gemütlich': Leben in der Dreieckssiedlung," in Stadt Recklinghausen (ed.), *Hochlarmarker Lesebuch*, 236.

56 Meyer-Renschhausen, *Energiepolitik in der BRD*, 28–9; J. Clark, *The Political Economy of World Energy*, 28, 62; Burckhardt, *Der Energiemarkt in Europa*, 47.

57 Schwartz Cowan, *More Work for Mother*, 196.

58 "Niederschrift über die Sitzung im Bundeskanzleramt am 6. August 1958," in Martiny and Schneider, eds., *Deutsche Energiepolitik seit 1945*, dokument 32.

59 Spiegelberg, *Energiemarkt im Wandel*, 30.

60 Housewife, born 1937, "Ich muß immer etwas zu tun haben," in Stadt Recklinghausen, ed., *Hochlarmarker Lesebuch*, 241.

61 Housewife, born 1928, in Stadt Recklinghausen, ed., *Hochlarmarker Lesebuch*, 235.

62 See also Yong-Sook, "Just a Housewife?"

63 "Werbespots: Kommissar Gluto," 4 mins., colour, ca. 1972–1974, Montanhistorisches Dokumentationszentrum (montan.dok), Deutsches Bergbau-Museum, Bochum.

64 Ludwig Windstosser, "Werbeaufnahme (Ruhrkohle)," in Schneider, ed., *Als der Himmel blau wurde*, photographs 299, 322.

65 Germany was one of the pioneering countries introducing district heating in the 1920s. Coal gas was distributed through gas pipelines in the Ruhr area to provide heating. For example, Ruhrgas AG was founded in 1926 to distribute the coal gas which was produced in the coal-mining coking plants. During the early 1960s, 90 per cent of its gas was still coke-derived, but by

1970 natural gas already comprised 70 per cent of its deliveries. For a history of Ruhrgas, see Bleidick, *Die Ruhrgas 1926 bis 2013*.

66   Abelshauser, *Der Ruhrkohlenbergbau seit 1945*, 89–93.

67   Brüggemeier, "A Nature Fit for Industry," 37. See also Nonn, "Vom Naturschutz zum Umweltschutz."

68   Brüggemeier, "A Nature Fit for Industry," 37.

69   Brüggemeier and Rommelspacher, *Blauer Himmel über der Ruhr*, 63.

70   Berta S., Hochfeld, "Soll in Hochfeld wohnen," *Westdeutsche Allgemeine Zeitung*, 24 and 25 September 1959, quoted in Brüggemeier and Rommelspacher, *Blauer Himmel über der Ruhr*, 206.

71   Mathilde E., Meiderich "Kinder sind in Gefahr," *Westdeutsche Allgemeine Zeitung*, 24 and 25 September 1959, quoted in Brüggemeier and Rommelspacher, *Blauer Himmel über der Ruhr*.

72   For a reproduction of selected photographs cf. "Kohle und Kommerz," *Die Zeit*, 2 June 2014, https://www.zeit.de/wirtschaft/2014-05/fs-chargesheimer-ruhrgebiet.

73   Böll, "Im Ruhrgebiet," 361.

74   Dolata-Kreutzkamp, *Die deutsche Kohlenkrise im nationalen und transatlantischen Kontext*.

75   "Bergarbeiter-Demonstration: Beinahe militant," *Der Spiegel* 40, 30 September 1959, 12–14, https://magazin.spiegel.de/EpubDelivery/spiegel/pdf/42622758.

76   Schneider, ed., *Als der Himmel blau wurde*, 78–84.

77   Adam, Reinhold, "'Mittlerweile habe ich mich daran gewöhnt': Schwierigkeiten durch den frühen Ruhestand." See Stiftung Geschichte des Ruhrgebiets, "Menschen im Bergbau," accessed 30 November 2019, https://menschen-im-bergbau.de/menschen/reinhold-adam.

78   V. Hall, *Women at Work*, 44. See also "Fraueninitiative war die Schulung fürs Leben," *Rheinische Post*, 19 April 2014.

8

Anthropocene Women: Energy, Agency, and the Home in Twentieth-Century Britain

Vanessa Taylor

As a very new demonstrator in a very new Board, I felt strange and uncertain, because I had first to convince myself that the new ways were going to be better than the ones I ... had grown up with. Having done this, I had to pass on my conviction to other people. At times, it was not so easy, because tradition dies hard.[1]

Introduction

The opening quotation comes from Edna Petrie in 1963, looking back on her early days as an electricity demonstrator for the North of Scotland Hydro-Electric Board (NSHEB). Founded in 1943, the Hydro Board had a mission to provide electricity for all in the Scottish Highlands and break the dominance of coal- and peat-fired cooking, heating, and washing. Their efforts saw a rise from just under 2,000 farms and crofts electrified in 1948 to over 33,000 by 1963: around 85 per cent of such properties in the region.[2] By that point, over half of the board's customers were using electric cookers, a higher market share than in any other region in Britain.[3] Edna Petrie was one of hundreds of demonstrators and home advisors employed by electricity and gas boards in mid-twentieth-century Britain. In addition, there were thousands of women active in voluntary organizations sponsored by energy suppliers (at first private, then nationalized bodies from 1948): the Electrical Association for Women (established 1924), the Women's Gas Federation (1935), and the Women's Advisory Council on Solid Fuel (1943).[4] These women in Britain, and their counterparts in other industrialized countries, were indispensable to the new and expanding demand for electricity and gas in twentieth-century homes. This chapter examines the role of such women in light of what we now know about fossil fuels. Though electricity and gas were promoted as clean alternatives to domestic coal fires in postwar Britain, their use alongside oil is now recognized to be at the heart of the global climate crisis. The residential sector alone was estimated to be contributing 14 per cent of all UK greenhouse gas emissions in 2016.[5]

Now seems a good moment to highlight once more women's agency in the transition to energy-intensive domestic technologies of the past century. The concept of the Anthropocene – the image of anthropogenic transformation and environmental damage on a geological scale – has attracted criticism since its formulation in 2000 for its vision of species-level agency. A critical body of social science and humanities literature has pointed instead to the distinct forms of agency rooted in capitalist power structures, to the large corporations and main beneficiaries of environmental exploitation, and its vast inequalities.[6] This is a necessary corrective to narratives of undifferentiated human agents. But what is striking is how absent women are in these accounts. From Andreas Malm and Alf Hornborg's critique of the concept of the Anthropocene as obscuring the real historic culprits – a "clique of white British men" at the dawn of industrialization – to Jason Moore's "capitalocene" or Raewyn Connell's recent discussion of institutionalized "power-oriented masculinity" in the "sociocene," women's role in the unfolding catastrophe caused by our rising reliance on energy-hungry technologies and polluting systems is largely invisible.[7] Beside Moore's important parallel between women's unpaid work in the home and the role of value-less nature in capitalist exploitation, we also need to trace the ways in which women's expanded agency in many industrialized countries has entailed their increased culpability.[8] This chapter argues that women should be included more fully in this "man-made" crisis.

Producer-led visions do not engage directly with existing work on women's complex engagements with domestic technologies over the past 200 years.[9] Agency in domestic energy transitions was distributed across a wide, varied network of actors – from energy suppliers to housing reformers, appliance manufacturers, different household members, and the logics built into technologies.[10] For electricity to take root in every home in the industrialized world required the active engagement of millions of users: people resisted new energy-intensive technologies or used them on their own inconvenient terms; they also adopted and promoted them. We need to look at the drivers of cars as well as their manufacturers; at the consumers of electricity as well as the energy barons. The purpose of this chapter is not to measure women's culpability but to trace some historical forms of their agency in the diffusion of fossil fuel technologies, and to acknowledge their fully human role in the environmental damage resulting from this.

Specific cases reveal more than generalized statements, and the next
section explores questions of agency and constraint in the context of
early and mid-twentieth-century rural Scotland. This chapter argues that
the energy industry affected women in specific ways in Britain from the
1920s to the 1970s. Changes in household technologies transformed
the daily lives of women, men, and children, though not in all the ways
hoped for by energy-industry professionals. Political debates, popular
discourse, and marketing relating to new energy forms and appliances
sought to constitute women ideologically as a distinct category – the
domestic energy consumer in the home – ignoring the complexities
of energy use by women and men within and beyond the walls of
the home. At the same time, women's involvement in the industry
as professional demonstrators and voluntary promoters both was a
function of their limited employment opportunities and opened up
new avenues for them.

Women in Scotland in the Electrical Age

Edna Petrie's short 1963 account of her career in the Scottish Highlands
and Islands, written for *The Electrical Age* (published by the Electrical
Association for Women), offers a window onto women's highly active
but often ambiguous roles in the diffusion of energy-intensive domestic
technologies, and the constraints and opportunities those roles reflected
and created. The area where Petrie spent her life and career was no
ordinary regional electricity board. As a largely rural area with extremely
difficult terrain and distinctive ways of life, the North of Scotland demonstrates the challenges faced by electricity suppliers trying to expand rural
networks at their most acute. The long-term legacy of the Highland
Clearances and a strong policy position on stemming rural depopulation lent urgency to the mission of the pioneering NSHEB (established
in 1943), which helped to pave the way for electricity nationalization in
Britain. With its unique "social clause," requiring it to "have regard to
the social and economic betterment of sparsely populated areas,"[11] the
board prioritized affordable power for domestic consumers, and sought
to encourage small rural industries. Its founding director, Tom Johnston,
cited his international models as the Hydro-Electric Power Commission
of Ontario (1906) and the Tennessee Valley Authority (1933).[12] But not
everyone shared this vision of Highland Scotland, with its damming
of rivers and flooding of valleys. Major landowners, amenity bodies
such as the National Trust, salmon fisheries, tourism interests, and local
authorities had to be navigated and consulted.[13]

With the North of Scotland's 137 inhabited islands and miles of sparsely populated mountains on the mainland, a report in 1942 had concluded that most of the region would remain without electricity.[14] But the Hydro Board proved inventive, working with a complex mix of fuels and modes of generation. By the early 1950s, almost 13,000 premises were served by diesel generators, with groups of islands linked by submarine cable. At this point there were as many diesel stations as hydro-electric stations. A decade later, several islands were still without electricity, with up to 2,000 households being supplied with bottled gas by the electricity board. There were also steam-power generators at Aberdeen and Dundee (supplied from Scottish coalfields), an experimental station at Altnabreac in the far north east – where peat was used to fuel a 500-kilowatt gas turbine – and a 100-kilowatt windmill at Costa Head contributing to supply on Orkney's Mainland. The Highland Grid was eventually supplemented by atomic energy from Dounreay, built by the UK Atomic Energy Authority in the late 1950s, though this was also a major user of electricity. The separate North of Scotland and southern Scotland electricity boards exchanged electricity in response to fluctuating supply and demand across seasonal and daily cycles. In addition, Highlanders continued to fuel their own premises with wood and peat, and by 1952 they were still using 925,000 tons of coal a year in fireplaces and stoves, alongside paraffin lamps or gas for lighting.[15]

Edna Petrie's career with the Hydro Board spanned the key early developments of the Highland Grid, including its work with diesel, coal-fired steam, and atomic power, as well as hydro-electric power. Electricity promotion was not her first choice of career. She had wanted to learn German at school in wartime Orkney, but "teachers were scarce" and she had been advised to "try Domestic Science instead": this was no accident, of course, given the educational pathway of most 1940s schoolgirls. Without German, she said, and with a "continued interest in the domestic arts," she went to the Aberdeen School of Domestic Science and joined the Hydro Board in 1947 as its second demonstrator. By 1948 she and her colleague Lydia Scott were delivering cooking demonstrations to packed houses twice a week in the Kirkwall electricity showroom on Orkney, and travelling widely elsewhere. As she later recalled of her travels between islands and the mainland in a "small boat ... tossed in the fast running tides," "inconvenience [was] hardly the word to describe the thrill I got from these unusual journeys." With Petrie's roots in Orkney and Scott's on a neighbouring island, these "local ladies" were well-placed to do the hard work of eroding local traditions.[16]

By the early 1950s, Petrie was expanding her skills in Aberdeen on Scotland's northeastern coast – at that time, as she noted, "a 'gas' stronghold." The war had halted the progress of electric cookers in privately owned or rented homes, and up to 1950 the Corporation of Aberdeen had not installed a single electric cooker in its council housing. Many such urban areas were the scene of rivalries between electricity and gas providers hoping to monopolize services in expanding local authority housing. But by 1952, the Hydro Board had agreed to provide electricity to council premises for a flat rate, irrespective of whether gas was also provided, allowing tenants "freedom of choice" in their appliances.[17] Petrie was part of the vigorous sales campaign that followed. As she noted, the Hydro Board needed Aberdeen's domestic customers for their diversity of load (to aid the "load factor" of its power stations) and relied on urban areas to subsidize its uneconomic remote areas. The efforts paid off: electric cooking by Aberdeen's council tenants rose from 9 per cent in 1952 to 80 per cent in 1963 (though across Scotland by this point, 50 per cent of cooking was still done on gas, coal, or peat-fire stoves). After Aberdeen, Petrie transferred to the north coast for "domestic reasons," facing her family's home island of Orkney across the Pentland Firth, working in the "'Atomic' town of Thurso." With many of the residents there employees of the new Dounreay nuclear power station, this was a "small but very cosmopolitan" community, according to Petrie: people were "not unnaturally, very electrically minded," and many of their homes, "all-electric."[18]

Despite the electricity industry's rhetoric, it is often unclear who the intended beneficiaries of domestic electricity really were: the consumers or suppliers? Technology historian Thomas P. Hughes famously said of electricity turbines, that with their constant drive to achieve sufficient load and economies of scale, they were "in effect … supply in search of demand."[19] The case of rural electrification was more complex, however. The Scottish Highlands, with their plentiful rivers, were perfect for creating hydroelectricity but a particularly unpromising place to search for demand. The logic that might have dictated a focus on easy-to-reach urban consumers was subordinated to policy ideals about universal provision and raising standards of living in early and mid-twentieth-century Britain.[20] The promotion of electricity for all became attached to long-standing calls for housing reforms. At the same time, energy transitions shaped changing views on what was and was not "fit for habitation," as yesterday's luxuries became today's necessities. But it was not just housing that seemed in need of updating as energy services expanded. Industry and policy makers also targeted entrenched everyday practices and forms of self-reliance that were unremunerative (from a supplier's viewpoint) in order to create well-drilled, monetized energy consumers who could help "build" the load.[21]

Modes of domestic energy provision were entwined with a multitude of everyday routines, and gas and electricity suppliers were up against both deep-rooted traditions that "died hard," as Petrie said, and more ephemeral conveniences. In 1927, groups of women in parts of London, as dyed-in-the-wool gas users, were said to be in "revolt" against plans by the local authority to introduce electric cookers. In the 1930s, women in the Scottish coastal region of Fife locked their doors against the local gas provider who came to change their prepayment meters, because their existing meter settings ensured regular refunds and were being used as a handy savings bank.[22] Petrie's claim in 1963 that electric lighting "was one thing that needed no commendation" was borne out by other electricity suppliers reporting from rural areas in the 1940s and 1950s. Indeed, some suppliers complained that *only* lighting was being used in the countryside, in preference to the many other energy-intensive and load-building appliances that could prove so useful around the home and farm if only people would invest in them.[23]

The frequent gaps in priorities between energy suppliers and appliance manufacturers on one hand, and users on the other, has been the focus of much attention in the literature on women's approaches to domestic technologies. This attention has demonstrated the inadequacy of supply-focused accounts of energy transition, looking instead at processes of appraisal by women within the home. To what extent new energy-use technologies made sense to potential buyers in light of their cost, ease of access, and interaction with existing household technologies and routines were just some of the issues Ruth Schwartz Cowan examined through her concept of the "consumption junction," for example. This body of work has shown, firstly, how new energy forms and appliances were taken up in homes only where they were successfully "domesticated," and secondly, that women were key agents in this process: both women inside the home and female promoters possessing the necessary authority to persuade other women.[24] Structural constraints within and beyond the home that determined their freedom to select fuels or appliances included housing tenure and socio-economic status, and questions of who had the power to make decisions within the home, in a context in which women in the UK could not sign their own credit agreements until the late 1970s.[25]

To what extent "labour-saving" devices were *imposed* on women remains an important question, given women's varied circumstances. So too is the question of the effects of these devices on women's lives, despite Schwartz Cowan's persuasive argument that the "semi-industrialized" household led not to less "women's work" but to higher expectations of domestic performance.[26] There is a wealth of evidence that women in inter- and postwar Britain were living in cold, damp housing, were lacking in nutrition and health care, and had daily domestic burdens that made them ill. A 1930s study of British housewives carried out by the Women's Health Enquiry Committee indicates the way in which concerns about women's housing, domestic labour, access to networked water and energy services, and health became entwined at this time. The study stated that "the present difficulties of water, lighting and heating should no longer be tolerated, even for old cottages," adding – as was commonly assumed – that if these problems were removed, "one of the causes of rural depopulation will disappear."[27]

The concerns of such studies have much in common with the more earnest discussions that were taking place at the same time about how electricity could make life easier for working-class women. There seems little doubt that Petrie was hoping for a better life for the people among whom she lived and worked – as well as improvements on the "paraffin and tilley-lamps" that could take "upwards of an hour preparing"; the solid fuel stoves used for cooking, washing, and heating; the cold bedrooms with their "registered grates" that were hardly ever used; and laundry day, "a major event" involving tubs of heated water, flat- and box-irons, a gap in the rain, and "the wind in the right direction." Electricity, Petrie claimed, was "now a life-line throughout Scotland, and especially so in the scattered countryside I've worked in"; it had "made life easier for them in every way."[28] Living in scenic Ross-shire (northwest Highlands) by 1963, she reported that the expanding tourist industry here had

> Raised the standard of living, and in the most unlikely places the visitor from the South finds washing machines and spin-driers, tumble-driers and rotary ironers. These things are necessities where there is such a heavy rainfall, and where there is no commercial laundry within 100 miles.[29]

The groundswell of women's housing and welfare groups working to raise living standards, alongside such energy promotion work, provided a vehicle for women's expanding citizenship during these years.[30] Though they did not all share the same vision of the future, the need for rural Britain to embrace modernity was one of the commonalities.

The phenomenon of the "blackhouse" shows how standards of living, modes of energy provision, and moral standards were frequently bundled together. Some of these blackhouses were still inhabited in remote islands in the 1940s. Here, as Petrie said, "[t]he cooking-pot was hung on a chain suspended from the roof, a fire lit under it, usually of peat, and the smoke found its way out of the room as best it could."[31] By 1963 these houses had long symbolized all that was wrong in rural housing in Scotland. Considered as hotspots for fever (for typhoid and typhus, especially), there were also persistent moral and cultural concerns about these one- and two-roomed dwellings shared between household members of different sexes and different generations, and sometimes between people and animals.[32] Electricity seemed to provide a moral as well as practical advance on this way of life. As the *Aberdeen Journal* put it in 1927, the "chimney-less house," though rare, was as "an advanced example of all the 'Electric House' and its advocates are out to banish, dirt and disease being among the foremost evils." Shiny new standards of lifestyles built on a US model cast shadows that fell not just on the blackhouses but on homes and their residents right across Britain. As the reporter continued:

> Though women interested in science, domestic and otherwise, and hygiene are among the prime-movers in the campaign to put the coal-and-gas ridden houses of Britain on the same level as the clean, cosy, and easy homes of America, the final success or failure of the movement lies with the woman in the house.[33]

Some women were at the vanguard of this campaign, but many other women themselves required an upgrade.

Forms of rural self-reliance on the part of both men and women were targeted during the interwar and postwar years, by the Hydro Board, by policy makers, and by various forms of propaganda whose ultimate source is not clear. The self-reliant practices of rural Scotland did not fit the model of paying energy consumers that was required by networked services. Like the blackhouses, the long-standing practice of collecting free peat from local peat bogs for domestic fuel, carried out by women and men, seemed to be a problem that needed solving. This kind of cash-free productive work was invisible in talk of women as primarily consumers: a false image in a context in which women frequently worked alongside their husbands and older children in interconnected activities such as fuel gathering and animal husbandry. A 1924 *Illustrated London News* feature suggests that this was partly a question of what work was deemed appropriate to different genders. While the crofter man and boy bearing driftwood represented a "sturdy race … that gave many sons in the war," the fuel gathering of women was a Highland tragedy: it was "a common sight to see women and children staggering home with great bundles of sticks on their backs."[34] There are indeed many photographs of women from rural Scotland at this time, digging

out peat or carrying it home on their backs, but in many cases knitting
phlegmatically as they went. In a report to the 1950 Ridley Committee,
the Scottish Hydro Board commended its own experimental peat-fired
gas turbine at Caithness as more cost effective than the individual efforts
of Highlanders.[35]

By the 1940s, policy makers were trying hard to find out what the
people of the Highlands wanted and what could help to stem depopulation, commissioning a series of social surveys that built on rural housing
reports.[36] But the results were ambiguous. A 1949 social survey on rural
depopulation in Scotland considered the conditions of remote locations
and out-migration. The answers that were gathered suggested a range of
issues, from dissatisfaction with the absence of electricity, gas, or water,
to the lack of good jobs or wanting a better social life. A small-scale
postal survey that disaggregated women and men did not suggest that
women were more concerned with poor housing than men. Contrary
to expectations, the wider survey found that people with access to better
facilities in their houses were *more*, not less, likely to migrate to towns.
Employment prospects were the key driver for migration, but, of those
wanting to move, "their households were, more often than the average,
enjoying a main water supply and main drainage, and were more often
provided with gas, electricity, telephone and refuse services."[37] The priorities of the more remote rural inhabitants remained obscure.

Despite the lack of clear evidence for what women (or men) wanted,
there were frequent generalized discussions about the housewife as a
unitary consumer, who could best be understood by those who shared
this consumer's viewpoint. Women had, as the *Aberdeen Journal* put it in
1927, "a clear part in the matter in advisory and utilitarian capacities."[38] A
decade later, Phyllis Thompson, writing in the Electrical Association for
Women's magazine also commended a "feminine regard for the more
practical details of housing," which, she noted, had led women "to urge
the electrification of the homes of the working class at a reasonable
cost." She highlighted the benefit of having "the woman's viewpoint"
represented on local authorities' electricity committees (in the era of
municipal electricity boards).[39]

Consulting Women

The confident statements about women's viewpoints, like those about
the benefits of modernity, were a convenience and rhetorical strategy
more than a description of reality. People knew that in fact women did
not all share the same viewpoint, nor were their views confined to purely

practical matters. The multiplicity of women's viewpoints was evident in highly political debates over energy nationalization in the late 1940s – in the debate, for example, over whether the British Housewives' League (formed circa 1945, and opponents of nationalization) should be represented on the Electricity and Gas Consultative Councils and Domestic Coal Consumers' Council: a Conservative move resisted by Labour.[40] The consultative councils of the regional electricity and gas boards were intended as a bridge between suppliers and a range of industrial, commercial, and domestic consumers, landowners, trade unions, and other interests. Though their real power is debateable,[41] the variety of organizations that ended up on them demonstrates a recognition that women's representation required a range of viewpoints. From the 1940s to the 1970s, these included the Electrical Association for Women, the Women's Gas Federation, and the Women's Advisory Council on Solid Fuel for their respective sectors, but also the Co-operative Women's Guild, the National Council of Women of Great Britain, the National Federation of Women's Institutes, the National Joint Committee of Working Women's Organisations, the Women's Royal Voluntary Service, the Scottish Women's Rural Institutes, the Scottish Women's Group of Public Welfare, and the Union of Jute and Flax Workers in the 1950s and 1960s (a Dundee-based textile union with a large female membership).[42] There were also individual women (often local authority representatives) on the district-level consultative committees reporting to the regional level. That no woman sat on the NSHEB itself between 1948 and 1972 suggests the limits of women's power in this sector. The board remained a stronghold of nine men, flanked by technical advisors and two influential committees, again all male with the exception of local grandee the Honorable Lady MacGregor of MacGregor OBE, a fixture on the Amenity Committee.[43]

The fortunes of the energy sector in Britain rose and fell between the 1920s and 1970s; ultimately their promotion efforts paid off, though the "solid [fuel] women" were fighting a losing battle by the mid-1950s, with the onset of clean air legislation.[44] Through these years, women's groups were vital in helping women to expand their appreciation of electricity's potential and in keeping gas use buoyed up against its younger competitor. The strategy of the energy industries in providing an ever-expanding array of domestic technologies for women chimes with Schwartz Cowan's argument about ever higher expectations of women's domestic performance. The Women's Gas Federation's "Brides-to-Be" classes and "Young Homemakers" wing aimed to ensure that the next generation growing up in the 1950s and 1960s would continue an allegiance to gas cooking and to rising domestic standards.[45] Women's work in the energy sector into the 1980s continued to reflect their constrained social and economic status. Their voluntary contributions of free time in this field were closely tied to their primary domestic roles and widespread exclusion from the labour market. It is also significant that Edna Petrie and many other women employed in this sector up to the early 1950s were unmarried, at a time when the marriage bar was being eroded only slowly.[46]

But the constraining context of women's engagement with the energy sector was not the whole story. Some saw the potential for labour-saving appliances in freeing women from the domestic sphere. The pioneering engineer Caroline Haslett, a founder of the Women's Engineering Society and Electrical Association for Women, was in 1947 already advocating for electricity as a way of allowing working-class women more time to do paid work, as "a valuable asset to the nation."[47] As a female engineer, Haslett (also unmarried) was atypical within the industry, but the sector offered a wide range of women – from engineers to domestic scientists and voluntary promoters – an expanding field of opportunities. The increasing professionalization of demonstrators and home service advisors during the 1950s is indicated by the Association of Home Economists (formed in 1954). In the 1960s and 1970s, as women became more difficult to find at home, home visits declined. Such face-to-face interaction became less valuable to the industry over time, though women continued to work as demonstrators in retail outlets and as freelancers, including paid work in print journalism and tele-vision demonstration.[48] By the 1980s, electricity had been successfully established. Annual sales of electricity per head of the British population rose from 442 kilowatt hours in 1938 to 4,137 kilowatt hours by 1980. Electricity sales in domestic residences and farms rose from 5,361 giga-watt hours in 1938 to 87,907 gigawatt hours in 1980.[49] Gas, in decline by the 1950s, had received a great boost with the discovery of North Sea gas (and oil fields), which enabled the roll-out of "high-speed gas" in the late 1960s and 1970s.

Women within the voluntary energy promotion organizations – often married – valued their expanded opportunities highly. In a 2016 interview, a former Women's Gas Federation (WGF) representative for Eastern England recalled the first time that she ever flew, when British Gas arranged for a helicopter trip for WGF council members to see the first of the North Sea gas rigs off the Norfolk coast at Bacton around the late 1960s: "I was petrified. I thought, 'I'm going to.' So I did."[50] "It was always about social friendship," she said: "I'm sure my members would say it enriched their lives, and it opened many doors."[51] While there were undoubtedly tensions between different groups – for example, between male (and some female) professionals and volunteers – new social communities were forged. As a former national chairman of the WGF recently told me:

> we had a couple of members whose sisters were home service advisors for an electricity company. We were all sort of friends together ... we wanted everybody to have a better life and better opportunities.[52]

WGF members launched campaigns on issues they cared about in the 1980s, including promoting equal pay, care for those with HIV, and tests for cervical cancer.[53] Elspeth Howe, the WGF president in the 1980s, was earlier deputy chair of the Equal Opportunities Commission and active in the campaign for women's right to sign their own credit agreements (for example, to buy appliances on credit cards or using hire purchase arrangements).[54] When the women's voluntary organizations ceased to be funded by the newly privatized energy sector in the early 1990s, many of the WGF local branches continued to meet as social groups, and are still active today as "CAMEO" clubs.[55] Through these activities women made the most of their opportunities, like men. It was never just about the energy.

Women did sufficiently extricate themselves from hard physical labour in the home to enter paid work, while bearing the dual burden of child-rearing. Women also chose not to become wives and mothers in unprecedented numbers. Following an overall twentieth-century trend, the female workforce in the UK rose from just over seven million to nearly twenty-three million between 1950 and 1990, while the participation of older, married women and mothers in the labour force expanded significantly. By 1995, nearly 67 per cent of women aged between fifteen and sixty-four were working, though the majority of working mothers were in part-time work, and women were still concentrated in a narrower range of sectors than men, and in the lower-paid jobs.[56] With rising living standards came rising expectations of household comfort, as the energy industry and policy makers had always wanted. Lack of central heating, the norm in the UK in the 1940s, became a measure of deprivation.[57] Despite post-war aspirations for universal provision, not all inhabitants of the UK have been equal beneficiaries of this rising affluence. In Scotland, nearly 25 per cent of households were considered to be in conditions of fuel poverty in 2017. Fuel poverty was higher among rural residents, still reliant on oil, more highly priced rather than gas. Remote rural fuel poverty was estimated at around 59 per cent in 2017.[58]

This is a story of women's role in achieving conditions of rising affluence for themselves and wider society, though the benefits have been unevenly distributed and have come at an environmental cost. As geographers Chris Gibson et al. put it: "The strongest predictor of carbon footprint/greenhouse gas emissions is affluence, at both the macro and the household scale."[59]

Conclusions

This chapter has traced some of the ways in which women's role in the energy sector – that of women such as Edna Petrie and others like her – was predicated on a gender hierarchy and ideologies of women's domestic role. The ideological framing of women as primary domestic energy consumers suppressed both their involvement in productive roles, within and beyond the home, and those forms of self-reliance in energy provision that seemed inimical to modern life and to appropriate gendered behaviour. Weaving women into the project of growing energy consumption was at the heart of the sector's domestic energy aims in the twentieth century. Many women engaged in this process with enthusiasm, if not on an equal footing with the men in charge of the industries, and with some ambivalence (as neither truly "domestic" women, nor fully equivalent to men). This was part of the route to women's emancipation from the home and the increasingly indus- trialized affluence of women, as well as men. Today in Britain, with the increased equality of women and men, we flicker in and out of our sexed and gendered identities as we move through our daily lives. We are also still framed by forms of distributed agency over which we have limited control, but British women today have a higher carbon footprint, partly as a result of pathways forged by the women who came before them. Globally, many women are very far from equivalent either to men or to women in affluent industrial nations.

The path to affluence and environmental degradation was paved with good intentions, as well as with relationships of exploitation. Can we envisage a different future for people that does not tread the same path, especially as global warming is widening environmental and economic disparities globally? While women in their domestic roles have – like men – exploited cheap nature, the idea of "women" as being in some ways outside of capitalism nevertheless still acts as a kind of cultural resource, similar to Nancy Fraser's (2014) "reservoirs of 'non-economic' normativity."[60] This idea holds out the promise of doing things differ- ently in the future, but it is far from clear what this means. If we want to realize this potential, we must not forget the many ways in which women have also always been part of the problem.

Notes

The author thanks the editors of this volume, colleagues in the Material Cultures of Energy project (AHRC grant AH/K006088/1), the Rachel Carson Center for a writing fellowship, and the women who shared with me their memories of energy promotional work.

1   Petrie, "A Demonstrator's Work in the North of Scotland," 497.
2   North of Scotland Hydro-Electric Board (NSHEB), *Annual Report and Accounts 1963*, 20.
3   Ibid., 19.
4   Pursell, "Domesticating Modernity," 47–67; Clendinning, *Demons of Domesticity*; Gooday, *Domesticating Electricity*.
5   Department for Business, Energy and Industrial Strategy. *2016 UK Greenhouse Gas Emissions, Final Figures*, 24–5. This is in addition to emissions arising from the generation of electricity *used* in the home.
6   Crutzen and Stoermer, "The Anthropocene," 17–18; Chakrabarty, "The Climate of History," 197–222. Cf. McAfee, "The Politics of Nature in the Anthropocene," 65–72; Meyer, "Politics In – but Not Of – the Anthropocene," 47–51; T. Cooper, "Why We Still Need a Human History in the Anthropocene." For existing connections, see, for example, Sandwell, "Pedagogies of the Unimpressed," 36–59; MacGregor, "Go Ask 'Gladys'"; Otter, et al., "Roundtable: The Anthropocene in British History," 584, 591.
7   Malm and Hornborg, "The Geology of Mankind?," 64; Malm, *Fossil Capital*; Moore, ed., *Anthropocene or Capitalocene?*; Moore, "The End of Cheap Nature," 1–31; Connell, "Foreword: Masculinities in the Sociocene," 6.
8   See Moore, "The End of Cheap Nature," 288, 301–2; also Fraser, "Behind Marx's Hidden Abode," 55–72. But on the capacity of liberal feminist gender equality to produce "ruling-class women", see Arruzza, Bhattacharya, and Fraser, *Feminism for the 99%*, 53.
9   See note 23.
10 On distributed agency, see Wilhite, "Energy Consumption As Cultural Practice," 60–72.
11 *Select Committee on Nationalised Industries*, 482; Chick, "Time, Water and Capital," 29–55.
12 Johnston, T., *Memories*, 181; "Hydro-Electric Development (Scotland) Bill."
13 Smout, *Nature Contested*.
14 SDD, *Electricity in Scotland*, 155; Scottish Office, *Report of the Committee on Hydro-Electric Development in Scotland*, 47.
15 For diesel, see SDD, *Electricity in Scotland*, 147, 156–62. See also NSHEB, *Annual Report 1952*, Appendix 7, pages 3, 4, 5, 11, 15; NSHEB, *Annual Report 1953*, 14, 16–17, map; NSHEB, *Annual Report 1972*, 3, 4, 9–11; and Payne, *The Hydro*, tables 24, 25.

16   Petrie, "A Demonstrator's Work in the North of Scotland," 496–7. See also
     "Electrical Cooking," *Orkney Herald, and Weekly Advertiser and Gazette
     for the Orkney & Zetland Islands*, 6 April 1948, 4; "By Kirsteen," *Aberdeen
     Evening Express*, 1 April 1952, 4; and "Sealing the Cookers," *Aberdeen Evening
     Express*, 16 December 1954, 11. Petrie came from Tankerness; Lydia Smith
     from Shapinsay.
17   Petrie, "A Demonstrator's Work in the North of Scotland," 497. See also
     "Smaller Electricity Bills," *Aberdeen Journal*, 2 December 1942, 4; "Electricity
     or Gas," *Scotsman*, 20 May 1950, 5; "Gas Board Will Seek Conference,"
     Aberdeen Evening Express, 9 April 1952, 5; "Gas V. Electricity Grouse: Tenants
     Don't Get Free Choice," *Dundee Courier*, 10 April 1952, 2; and NSHEB, *Annual
     Report 1953*, 23. For local authorities as "multiple" or "proxy consumers" of
     appliances, see Schwartz Cowan, "The Consumption Junction," 266, 269, 270;
     and Trentmann and Carlsson-Hyslop, "The Evolution of Energy Demand in
     Britain," 807–39.
18   Petrie, "A Demonstrator's Work in the North of Scotland," 497–8. See also
     NSHEB, *Annual Report 1972*, 3, 13, map ix. For load factor, see Hughes,
     Networks of Power, 217–21; and Roberts, "Electrification," 68–112.
19   Hughes, *Networks of Power*, 364.
20   Brassley, Burchardt, and Sayer, "Conclusion: Electricity, Rurality and
     Modernity," 221–45.
21   Savage, *Rural Housing*; Thresh, *The Housing of the Agricultural Labourer*;
     Ministry of Health, *Rural Housing*, 15, 41; Smith, M., *A Guide to Housing*;
     Ambrose, *The Quiet Revolution*, 190–1; SDD, *Electricity in Scotland*, 151; Gillott,
     "Domestic Load Building," 197–202.
22   "Electrifying the Home," *Aberdeen Journal*, 26 February 1927, 6; "Fife
     Housewives' Protest: Gas Meter Alterations Resented," *Dundee Courier*,
     12 December 1934, 7.
23   Petrie, "A Demonstrator's Work in the North of Scotland," 496, 497. See, for
     example, South West Electricity Consultative Council Chairman in "Women
     Are Leaving Countryside: Current Is their Great Need," *Western Daily Press*,
     6 July 1949, 6.
24   Schwartz Cowan, "The Consumption Junction," 253–72; Parr, "What Makes
     Washday Less Blue?," 153–86; Sandwell, "Pedagogies of the Unimpressed";
     Sandwell, "How Households Shape Energy Transitions," 23–30; Oldenziel,
     "Man the Maker," 128–48; Gooday, *Domesticating Electricity*.
25   Taylor and Chappells, "What Consumers in the Past Tell Us About Future
     Energyscapes," 11–21. For credit, see 1974 Consumer Credit Act, 1975 Sex
     Discrimination Act.
26   Schwartz Cowan, *More Work for Mother*.
27   The Women's Health Enquiry Committee, established in 1933, included
     women from the following organizations: the Council of Scientific

Management in the Home (National Council of Women), the Midwives Institute, the National Council for Equal Citizenship, the National Union of Townswomen's Guilds, the North Kensington Women's Welfare Venture, the Standing Joint Committee of Industrial Women's Organisations, the Women Public Heath Officers' Association, the Women's Cooperative Guild, and the Women's National Liberal Federation. For the quotation, see Spring Rice, *Working-Class Wives*, 196–7.

28 Petrie, "A Demonstrator's Work in the North of Scotland," 496, 498.

29 Ibid., 498.

30 Meller, "Women and Citizenship," 234–5; Llewellyn, "Designed by Women and Designing Women," 42–60; Hannam, "Women As Paid Organizers and Propagandists," 69–88; Cowman, "*From the Housewife's Point of View*," 352–83.

31 Petrie, "A Demonstrator's Work in the North of Scotland," 496.

32 Blackhouses varied but tended to be small dwellings with dry-stone walls, a turf or thatched roof, a central hearth and no chimney. The interior space was largely shared, though later blackhouses could have partitions made of turf or planks. They were found especially in the Scottish islands and mainland coastal areas. Housing reformers and some landlords discouraged them from the late nineteenth century onwards. Royal Commission on Housing of Working Classes, *Second Report (Scotland)*, 20, 910–19. See also "The Black Houses in Harris," *Scotsman*, 8 May 1896, 6; "'Bundling' System in the Lews," *North British Daily Mail*, 5 June 1900, 3; and "Highland Needs," *Aberdeen Press and Journal*, 14 June 1919, 4. See also Smith, "The Housing of the Scottish Farm Servant"; as well as the discussion of the "bothies"/"chaumers" of unmarried male workers, in Scottish Housing Advisory Committee (SHAC), *Report on Rural Housing in Scotland*, 103.

33 "Electrifying the Home," *Aberdeen Journal*, 26 February 1927, 6.

34 "The Plight of the Highland Crofter: Starvation in the Hebrides," *Illustrated London News*, 8 March 1924, 10.

35 NSHEB, *Annual Report 1952*, appendix VII, 4. See also British Pathé, "Peat"; British Pathé, "Peat Utilisation Scheme"; Scottish Film Council, et al., *Crofter Boy*. For photographs, see, for example, "The 'Buts' of Lewis," *Picture Post*, 28 May 1955 (Malcolm Dunbar photo); and "The Crofters' Isle," *Picture Post*, 3 September 1955 (Bert Hardy photo) (Hulton Archive/Getty Images). See also discussion in Fleetwood, "The Electrification of Scotland," 79–80.

36 SHAC, *Report on Rural Housing in Scotland*, 90; SHAC, *Distribution of New Houses in Scotland*, for example, 92–6, 205; Box and Thomas, "The Wartime Social Survey," 164–7.

37 Hutchinson, *The Social Survey*, 9, 27–8, tables 21, 22, 23, 32, 33, appendix II. The provision of electricity to rural areas did not succeed in stemming the depopulation of the Highlands. See also Payne, *The Hydro*, 206.

38 "Electrifying the Home," *Aberdeen Journal*, 26 February 1927, 6.

39 Thompson, "Women on Electricity Committees," 334–5. For women's authority in reaching female consumers, see Gooday, *Domesticating Electricity*, 33–4.

40 "Clause 7 (Consultative Councils)," HC Deb 24 June 1947. For the Housewives' League, see McCarty, "Irene May Lovelock (1896–1974)," in *Dictionary of National Biography*.

41 Tivey, "Quasi-Government for Consumers," 137–51; Hilton, *Consumerism in Twentieth-Century Britain*, 147–9.

42 NSHEB, *Annual Report 1953*, appendix VI, Consultative Council Report; NSHEB, *Annual Report 1963*, appendix V, Consultative Council Report; NSHEB, *Annual Report 1972*, appendix VIII, Consultative Council Report; "Written Answers," HC Deb 19 January 1968.

43 NSHEB, *Annual Report 1953*, 4–5; NSHEB, *Annual Report 1963*, vi–viii; NSHEB, *Annual Report 1972*, v–vii. See, however, the call for woman on the board from George Thomson, Labour MP for Dundee East, and Dame Irene Ward, Tynemouth MP: "North of Scotland Hydroelectric Board (Report and Accounts)," HC Deb 27 January 1958.

44 Mosley, "Clearing the Air."

45 These classes were renamed "Teens and Twenties" in the 1960s. See Taylor, "Material Cultures of Energy," interview no. 4, 10 March 2016.

46 See, for example, Sex Disqualification (Removal) Act 1919 and "Civil Service Marriage Bar (Abolition)," HC Deb 15 October 1946. After the Home Civil Service marriage bar was lifted in 1946, local authority members across Scotland and elsewhere worked to continue the ban. See "Married Women: Employment in Municipal Service," *Scotsman*, 16 January 1946, 6; "Employment of Married Women: Dundee Committee's Discussion," *Scotsman*, 27 December 1946, 3; and "Time Limit on Civic Jobs for Mrs," *Dundee Courier*, 23 April 1952, 2.

47 Haslett, "Electricity in the Home," 651–2. For Haslett, see also Gooday in this volume.

48 MCE Oral History Interview, no. 4, 10 March 2016 (VT). For the professionalization of home economists, see Pratt, "Home Economics Subject Development in the Context of Secondary Education," 33, 93–4, 108–10.

49 Hannah, *Engineers, Managers, and Politicians*, table A2.

50 MCE Oral History Interview, no. 3, 2 March 2016 (VT). Likely to have happened around 1968 to 1971.

51 Ibid.

52 MCE Oral History Interview, no. 5, 10 March 2016 (VT).

53 Ibid.

54 See, for example, "City Notes: Lady Howe Takes Shopkeepers to Task," *Birmingham Daily Post*, 16 March 1978, 8.

55 For example, Coedpoeth CAMEO (Come and Meet Each Other), North Wales.
    MCE Oral History Interview, no. 8, 12 March 2016 (VT).
56 Walsh and Wrigley, "Womanpower," 2, table 1. See also Baroness Elliot of
    Harwood in "Women in Industry and the Home," HL Deb 19 June 1963.
57 Department for Communities and Local Government, *The English Indices of
    Deprivation 2010*, 15–16.
58 Fuel poverty is defined here as: "to maintain a satisfactory heating regime, it
    would be required to spend more than 10% of its income on all household
    fuel use." Scottish Government (Housing and Social Justice Directorate),
    "Scottish House Condition Survey: 2017 Key Findings" (4 December
    2018), Sections 4, Nos. 137, 145, 180–1, https://www.gov.scot/publications/
    scottish-house-condition-survey-2017-key-findings/pages/6 (accessed
    10 November 2019).
59 Gibson, et al., "Climate Change and Household Dynamics," 4.
60 Fraser, "Behind Marx's Hidden Abode," 69.

Bibliography

Archives

AGEB (AG Energiebilanzen e.V.). "Primärenergieverbrauch: Alte Bundesländer." Accessed 13 December 2019. https://ag-energiebilanzen. de/12-0-Zeitreihen-bis-1989.html.
– "Struktur des Energieverbrauchs: Alte Bundesländer." Accessed 13 December 2019. https://ag-energiebilanzen.de/12-0-Zeitreihen-bis-1989.html.
Behind the Kitchen Door Project. File 4008:0014. British Columbia Archives, Victoria, BC.
British Columbia Electric Co. Ltd, *A Chat on Electricity* (c. 1904). Royal British Columbia Archives and Library, NWP 971.91, B863. Victoria, British Columbia.
Bunreacht na hÉireann (Constitution of Ireland). Dublin: Stationary Office, 2015.
Central Statistics Office. "That Was Then, This Is Now: Change in Ireland, 1949–1999; A Publication to Mark the 50th Anniversary of the Central Statistics Office." Dublin: Stationary Office, 2000. Https://www.cso. ie/en/media/csoie/releasespublications/documents/otherreleases/ thatwasthenthisisnow.pdf.
Digital Repository of Ireland. Life Histories and Social Change Collection. https://repository.dri.ie/catalog/9593xp97w.
ESB Archives. Power Station Portfolio. https://esbarchives.ie/portfolio.
– National Breadbaking Championships. 2019.
HCHC, "Sir George Beilby – 1850–1924," *Proceedings of the Royal Society of London. Series A*, 109, no. 752, 1925, i–xxix.

Parliamentary Papers, 1854–55: Numerical List and General Index, Session
12 December 1854–14 August 1855. Account and Papers, vol. 57 (London,
House of Commons, 1855). "Mr Collier – Return of the Quantities of
Linseed, Corn, Hemp, Flax, or Tallow Imported during the Year 1854,
into the United Kingdom, from the Black Sea..." 9 March 1855. MERL
Pamphlet 4822 Box 1/11.
Proceedings of the Old Bailey, 1674–1913. Trial of Elizabeth Hallick
(t18300218-61), February 1830. www.oldbaileyonline.org.

Parliamentary Papers, Annual Reports, and Official Papers

House of Commons of the United Kingdom. Civil Service Marriage Bar
(Abolition). HC Deb 15 October 1946 vol 427 cc794–6. Accessed 9 July
2020. https://api.parliament.uk/historic-hansard/commons/1946/oct/15/
civil-service-marriage-bar-abolition.
– Clause 7 (Consultative Councils). HC Deb 24 June 1947 vol 439 cc231–
83. Accessed 9 July 2020. https://api.parliament.uk/historic-hansard/
commons/1947/jun/24/clause-7-consultative-councils.
– Hydro-Electric Development (Scotland) Bill. Order for Second Reading.
HC Deb 24 February 1943 vol 387 cc182–260, 233. Accessed 9 July 2018.
https://api.parliament.uk/historic-hansard/commons/1943/feb/24/hydro-
electric-development-scotland-bill.
– North of Scotland Hydroelectric Board (Report and Accounts). HC Deb 27
January 1958 vol 581 c103–66, 145. Accessed 3 November 2019. https://
api.parliament.uk/historic-hansard/commons/1958/jan/27/north-of-
scotland-hydroelectric-board.
– Written Answers to Questions. HC Deb 19 January 1968 vol 756
cc702–4W. Accessed 3 November 2019. https://hansard.parliament.uk/
Commons/1968-01-19/debates/7328b6eb-95b5-4780-8d88-ab3fcfe5b6fa/
WrittenAnswers.
House of Lords of the United Kingdom. Women in Industry and the
Home. HL Deb 19 June 1963 vol 250 cc1298–337, 1,308–12. Accessed 4
November 2019. https://api.parliament.uk/historic-hansard/lords/1963/
jun/19/women-in-industry-and-the-home-1.

Debates

Hutchinson, Bertram. "The Social Survey: Depopulation and Rural Life in Scotland. A Summary Report of Three Inquiries for the Department of Health for Scotland in Parts of Rural Scotland as to the Causes of Rural Depopulation." London: Central Office of Information, 1949. National Archives, UK: RG23/136.

Ministry of Health. "Rural Housing: Third Report of the Rural Housing Sub-Committee of the Central Housing Advisory Committee." London: HMSO, 1944.

North of Scotland Hydro-Electric Board (NSHEB). "Annual Report and Accounts," 1952; 1953; 1963; 1972.

Royal Commission on Housing of Working Classes. "Second Report (Scotland), Minutes of Evidence, Appendix, Index," (C 4409), 1885.

Scottish Development Department (SDD). "Electricity in Scotland: Report of the Committee on the Generation and Distribution of Electricity in Scotland." Cmd 1859. Edinburgh: HMSO, 1962.

Scottish Housing Advisory Committee (SHAC). "Distribution of New Houses in Scotland." Cmd 6552. Edinburgh: HMSO, 1944.

– "Report on Rural Housing in Scotland." Cmd 5462. Edinburgh: HMSO, 1937.

Scottish Office. "Report of the (Cooper) Committee on Hydro-Electric Development in Scotland." Cmd 6406. 1942.

Select Committee on Nationalised Industries. "Report from the Select Committee on Nationalised Industries: The Electricity Supply Industry, Vol. I (Report and Proceedings)." HC236-I. London: HMSO, 1963.

Newspapers, Magazines, and Periodical Articles

Aberdeen Evening Express
Aberdeen Press and Journal
Art Journal
Birmingham Daily Post
Calgary Weekly Herald
Canadian Pharmaceutical Journal
Daily News
Derby Mercury
Dundee Courier
Edmonton Bulletin
Einheit: Organ der Industriegewerkschaft Bergbau und Energie
Electrical Review
Farmer's Magazine

Globe (Toronto, Ontario)
Illustrated London News (London, England)
Irish Times
Lamp: A Magazine Published in the Interest of the Employees of the Standard Oil Company
Liverpool Mercury
London Advertiser (London, Ontario)
Monetary Times, Trade Review and Insurance Chronicle
North British Daily Mail
Orkney Herald, and Weekly Advertiser and Gazette for the Orkney and Zetland Islands
Picture Post
REO *News*
Sanitary Journal
School Magazine
Scientific American
Scotsman
Times (London, England)
Wetaskiwin Timessanitary
Western Daily Press
Woman Engineer
Women's Penny Paper
World (Toronto, Ontario)

Websites

ALDI. "The National Brown Bread Baking Competition." https://www.aldi.ie/brown-bread-competition.
British Pathé. "Peat 1949." https://www.britishpathe.com/video/peat/query/peat.
British Pathé. "Peat Utilisation Scheme 1955." https://www.britishpathe.com/video/peat-utilisation-scheme/query/peat+scotland.
Irish Life and Lore. "Search: ESB." https://www.irishlifeandlore.com/product-tag/esb.
Scottish Film Council/Scottish Educational Film Association/Joint Production Committee of the Scottish Education. *Crofter Boy* (1955). https://www.bfi.org.uk/films-tv-people/4ce2b69e9c180.
SuperValu. "About Supervalu." https://supervalu.ie/about.

Interviews

Adam, Reinhold. "'Mittlerweile habe ich mich daran gewöhnt': Schwierigkeiten durch den frühen Ruhestand." *Menschen im Bergbau*. n.d. Accessed 30 November 2019. https://menschen-im-bergbau.de/menschen/ reinhold-adam.

Bergerhoff-Wodopia, Bärbel. "'Auch die Nachbarn waren alle Bergleute': Kindheitserinnerungen an ein Leben mit dem Bergbau." *Menschen im Bergbau*. n.d. Accessed 30 November 2019. https://menschen-im-bergbau. de/menschen/b.

O'Brien, Sorcha. "Electric Irish Homes: Rural Electrification, Domestic Products and Irish Women in the 1950s and 1960s." Kingston University and the National Museum of Ireland, 2016–2019. https:// electricirishhomes.org/about.

Sandwell, R.W. "Heat, Light and Work in the Canadian Home: A Social History of Energy, 1850–1950." Unpublished research project.

Stiftung Geschichte des Ruhrgebiets. "Menschen im Bergbau." Stiftung Geschichte des Ruhrgebiets and Deutschen Bergbau-Museum Bochum, 2018. https://menschen-im-bergbau.de.

Taylor, Vanessa. "Material Cultures of Energy (MCE)" oral history interviews. University of Greenwich, UK, 2016.

Published Texts

Abelshauser, Werner. *Der Ruhrkohlenbergbau seit 1945: Wiederaufbau, Krise, Anpassung*. Munich: Beck, 1984.

Accampo, Eleanor. *Industrialization, Family Life and Class Relations: Saint Chamond, 1815–1914*. Berkeley, CA: University of California Press, 1989.

Accum, Friedrich Christian. *A Practical Treatise on Gas-Light: Exhibiting a Summary Description of the Apparatus and Machinery Best Calculated for Illuminating Streets, Houses, and Manufactories, with Carburetted Hydrogen, or Coal-Gas: with Remarks on the Utility, Safety, and General Nature of This New Branch of Civil Economy*. London: R. Ackerman, 1815.

Adam, Reinhold. "'Mittlerweile habe ich mich daran gewöhnt': Schwierigkeiten durch den frühen Ruhestand." *Menschen im Bergbau*. Accessed 30 November 2019. https://menschen-im-bergbau.de/menschen/ reinhold-adam.

Adams, Annmarie. *Architecture in the Family Way: Doctors, Houses and Women, 1870–1900*. Montreal and Kingston: McGill Queen's University Press, 1996.

Adams, Carol J., and Lori Gruen. *Ecofeminism: Feminist Intersections with Other Animals and the Earth*. New York: Bloomsbury, 2014.

Adams, Sean Patrick. *Home Fires: How Americans Kept Warm in the Nineteenth Century.* Baltimore, MD: Johns Hopkins University Press, 2014.

Agarwal, Bina. "Does Women's Proportional Strength Affect Their Participation? Governing Local Forests in South Asia," *World Development* 38, no. 1 (2010): 98–112.

– "Re-sounding the Alert: Gender, Resources, and Community Action," *World Development* 25, no. 9 (1973): 1372–81.

Alglave, Émile, and J. Boulard. *La lumière électrique: Son histoire, sa production et son emploi...* Paris: Firmin-Didot, 1882.

Allen, Edith. *Mechanical Devices in the Home.* Peoria, IL: Manual Arts Press, 1922.

Allen, Michael Thad, ed. *Technologies of Power: Essays in Honor of Thomas Parke Hughes and Agatha Chipley Hughes.* Cambridge, MA: MIT Press, 2001.

Allen, Robert C. *The British Industrial Revolution in Global Perspective.* Cambridge, UK: Cambridge University Press, 2009.

– "The Shift to Coal." In *Energy Transitions in History: Global Cases of Continuity and Change,* vol. 2, edited by Richard Unger, 11–16. Munich: Rachel Carson Centre Perspectives, 2013.

Ambrose, Peter John. *The Quiet Revolution: Social Change in a Sussex Village 1871–1971.* London: Chatto and Windus/Sussex University Press, 1974.

Andrews, Thomas G. *Killing for Coal: America's Deadliest Labor War.* Cambridge, UK: Harvard University Press, 2008.

Appleyard, Rollo. *Charles Parsons: His Life and Work.* London: Constable and Co., 1933.

Arruzza, Cinzia, Tithi Bhattacharya, and Nancy Fraser. *Feminism for the 99%: A Manifesto* London: Verso, 2019.

Aufenanger, Jörg. "Der Familienskandal." *Gegenworte: Hefte für den Disput über Wissen* 29 (2013): 40–3.

Backhouse, Constance. *Petticoats and Prejudice: Women and the Law in Nineteenth Century Canada.* Toronto, ON: University of Toronto Press, 1991.

Baillargeon, Denyse. *Making Do: Women, Family and Home in Montreal During the Great Depression.* Translated by Yvonne Klein. Waterloo, ON: Wilfrid Laurier Press, 1999.

Baker, William C. *Artificial Warmth and Ventilation and the Common Modes by Which They Are Produced by Wm.C. Baker Practical Engineer in Warming and Ventilation,* revised edition. New York: J.F. Trow, 1860.

Baldry, G., *The Rabbit Skin Cap: A Tale of a Norfolk Countryman's Youth Written in his Old Age.* Edited by L.R. Haggard. London: Collins, (1939) 1950.

Barca, Stefania. "Energy, Property and the Industrial Revolution Narrative." *Ecological Economics* 70, no. 7 (2011): 1,309–15.

Baroth, Hans Dieter. *Aber es waren schöne Zeiten.* Cologne: Kiepenheuer and Witsch, 1978.

Barrett, Ross, and Daniel Worden, eds. *Oil Culture*. Minneapolis, MN: University of Minnesota Press, 2014.

Beaumont, Caitriona. "Women, Citizenship and Catholicism in the Irish Free State, 1922–1948." *Women's History Review* 6, no. 4 (1997): 563–85.

Beecher, Catharine, and Harriet Beecher Stowe. *The American Woman's Home*. New York: J.B. Ford and Company, 1869.

Beeton, [Isabella] Mrs. *Mrs. Beeton's Household Management: A Guide to Cookery in All Branches: Daily Duties, Menu Making, Mistress and Servant, Home Doctor, Hostess and Guest, Sick Nursing, Marketing, the Nursery, Trussing and Carving, Home Lawyer*. London: Ward, Lock, 1907.

Bell, Shannon Elizabeth, and Yvonne A. Braun. "Coal, Identity, and the Gendering of Environmental Justice Activism in Central Appalachia." *Gender and Society* 24, no. 6 (2010): 794–813.

Berg, Maxine. *The Age of Manufactures: Industry, Innovation and Work in Britain, 1700–1820*. London: Fontana Press, 1985.

– "What Difference Did Women's Work Make to the Industrial Revolution?" *History Workshop Journal* 35 (1993): 22–44.

– "Women's Work and Mechanization." In *The Historical Meaning of Work*, edited by Patrick Joyce. Cambridge, UK: Cambridge University Press, 1987.

Bergerhoff-Wodopia, Bärbel. "'Auch die Nachbarn waren alle Bergleute': Kindheitserinnerungen an ein Leben mit dem Bergbau." *Menschen im Bergbau*. Accessed 30 November 2019. https://menschen-im-bergbau.de/menschen/b.

Bernstein, Jennifer. "On Mother Earth and Earth Mothers: Why Environmental History Has a Gender Problem." *Breakthrough Journal* 7 (summer 2017). https://thebreakthrough.org/journal/issue-7/on-mother-earth-and-earth-mothers.

Beynon, Lillian K. "Women's Clubs: Their Nature and Purpose." *Report of the First Annual Convention of the Homemakers' Club of Saskatchewan – Held at Regina, Sask., Jan. 31, Feb. 1, 2, and 3, 1911*, 24–8. Saskatoon, SK: s.n, 1911. Accessed 14 July 2020. http://peel.library.ualberta.ca/bibliography/3663.html.

Black, Brian. *Petrolia: The Landscape of America's First Oil Boom*. Baltimore, MD: Johns Hopkins University Press, 2000.

Bleidick, Dietmar. *Die Ruhrgas 1926 bis 2013: Aufstieg und Ende eines Marktführers*. Berlin: Walter De Gruyter/Oldenbourg, 2018.

Böll, Heinrich. "Im Ruhrgebiet." In *Werke, Kölner Ausgabe*, vol. 10: 1965–1959. Cologne: Kiepenheuer and Witsch, 2005.

Bowen, Sarah, and Joslyn Sinnika Elliott. "Joy of Cooking?" *Contexts* 13, no. 3 (2014): 20–5.

Bowers, Brian. *Lengthening the Day: A History of Lighting Technology*. Oxford, UK: Oxford University Press, 1998.

Box, Kathleen, and Geoffrey Thomas. "The Wartime Social Survey." *Journal of the Royal Statistical Society* 107, no. 3/4 (1944): 164–7.

Boydston, Jeanne. *Home and Work: Housework, Wages and the Ideology of Labor in the Early Republic.* Oxford, UK: Oxford University Press, 1990.

– "To Earn Her Daily Bread: Housework in Antebellum Working-Class Subsistence." *Radical History Review* 35 (1986): 13–14.

Boym, Svetlana. *The Future of Nostalgia.* New York: Basic Books, 2001.

Bozzoli, Belinda. "Interviewing the Women of Phokeng." In *The Oral History Reader*, edited by Robert Perks and Alistair Thomson, 212–22. Abingdon, UK: Routledge, 2016.

Brassley, Paul, Jeremy Burchardt, and Karen Sayer. "Conclusion: Electricity, Rurality and Modernity." In *Transforming the Countryside: The Electrification of Rural Britain*, edited by Paul Brassley, Jeremy Burchardt, and Karen Sayer, 221–45. London: Routledge, 2017.

Brewer, Priscilla J. *From Fireplace to Cookstove: Technology and the Domestic Ideal in America.* Syracuse, NY: Syracuse University Press, 2000.

Briggs, Asa. *Victorian Things.* London: Batsford, 1988.

Brontë, Anne. *The Tenant of Wildfell Hall.* London: Penguin Random House, (1848) 1996.

Brown, Kate. *Plutopia: Nuclear Families, Atomic Cities, and the Great Soviet and American Plutonium Disasters.* Oxford, UK: Oxford University Press, 2013.

Brown, Terence. *Ireland: A Social and Cultural History, 1922–2002.* 2nd ed. London: Harper Perennial, 2004.

Browning, Robert. "Meeting at Night." *Aldine* 4, no. 2 (1871): 36. Accessed 15 July 2020. doi:10.2307/20635989.

Brüggemeier, Franz-Josef. "A Nature Fit for Industry: The Environmental History of the Ruhr Basin, 1840–1990." *Environmental History Review* 18, no. 1 (1994): 35–54.

Brüggemeier, Franz-Josef, and Thomas Rommelspacher. *Blauer Himmel über der Ruhr: Geschichte der Umwelt im Ruhrgebiet 1840–1990.* Essen, Germany: Klartext, 1992.

Burckhardt, Helmuth. *Der Energiemarkt in Europa: Energiewirtschaftliche und energiepolitische Beiträge zur Diskussion der Gegenwart.* Tübingen, Germany: Mohr, 1963.

Byatt, Ian. *British Electrical Industry: 1875–1914.* Oxford, UK: Clarendon Press, 1979.

Callen, Anthea, "Sexual Division of Labour in the Arts and Crafts Movement." In *A View from the Interior: Women and Design*, edited by Judy Attfield and Pat Kirkham, 148–62. London: Women's Press, 1989.

Campbell, Gary, and Laurajane Smith. "'Nostalgia for the Future': Memory, Nostalgia and the Politics of Class." *International Journal of Heritage Studies* 23, no. 7 (2017): 612–27.

Campbell, Gary, Laurajane Smith, and Margaret Wetherell. "Nostalgia and Heritage: Potentials, Mobilisations and Effects." *International Journal of Heritage Studies* 23, no. 7 (2017): 609–11.

Canel, Annie, Ruth Oldenziel, and Karin Zachmann. *Crossing Boundaries, Building Bridges: Comparing the History of Women Engineers, 1870s–1990s.* Amsterdam: Harwood Academic, 2000.

Cardia, Emanuela. "Household Technology: Was It the Engine of Liberation?" 2008 Meeting Papers 826, Society for Economic Dynamics, 2008. https://ideas.repec.org/p/red/sedoo8/826.html.

Carson, Rachel. *Silent Spring.* Greenwich, CT: Fawcett Publications, 1962.

Castaneda, Christopher. *Invisible Fuel: Manufactured and Natural Gas in America, 1800–2000.* New York: Twayne, 1999.

Cavert, William M. "Industrial Coal Consumption in Early Modern London." *Urban History* 44, no. 3 (2017): 424–43.

– *The Smoke of London: Energy and Environment in the Early Modern City.* Cambridge, UK: Cambridge University Press, 2016.

Chakrabarty, Dipesh. "The Climate of History: Four Theses." *Critical Inquiry* 35, no. 2 (winter 2009): 197–222.

Chambers's Information for the People 2, no. 91 (1842).

Chapman, Peter. *Fuel's Paradise: Energy Options for Britain.* London: Penguin Books, 1975.

Cherry, Deborah. *Beyond the Frame: Feminism and Visual Culture, Britain 1850–1900.* London: Routledge, 2000.

Chick, Martin. "Time, Water and Capital: The Unintended Contribution of the North of Scotland Hydro-Electric Board to the Application of Welfare Economics in Britain, 1943–1967." *Scottish Business and Industrial History* 25 (2009): 29–55.

Christian, David. *Maps of Time: An Introduction to Big History.* Berkeley, CA: University of California Press, 2004.

Cipolla, Carlo. *The Economic History of World Population.* Baltimore, MD: Penguin Books, 1962.

Clark, Anna. *The Struggle for the Breeches: Gender and the Making of the English Working Class.* Berkeley, CA: University of California Press, 1995.

Clark, John G. *The Political Economy of World Energy: A Twentieth-Century Perspective.* New York: Harvester Wheatsheaf, 1990.

Clark, Theodore Minot. *The Care of a House: A Volume of Suggestions to Householders, Housekeepers, Landlords, Tenants, Trustees and Others, for the Economical and Efficient Care of Dwelling-Houses.* New York: Macmillan, 1903.

Clear, Caitriona. *Women of the House: Women's Household Work in Ireland, 1926–1961: Discourses, Experiences, Memories.* Dublin: Irish Academic Press, 2000.

Clendinning, Anne. "'Deft Fingers' and 'Persuasive Eloquence': The 'Lady Demons' of the English Gas Industry." *Women's History Review* 9, no. 3 (2000): 501–37.

– *Demons of Domesticity: Women and the English Gas Industry, 1889–1939.* Hampshire, UK: Ashgate Publishing, 2004.

– "Gas and Water Feminism: Maud Adline Brereton and Edwardian Domestic Technology." *Canadian Journal of British History* 33 (1998): 1–24.

Clifford, Jim. "London's Soap Industry and the Development of Global Ghost Acres in the Nineteenth Century." *Environment and History* 26, no. 4 (2020).

Cobbett, William. *Cottage Economy: Containing Information Relative to the Brewing of Beer, Making of Bread, Keeping of Cows, Pigs, Bees, Ewes, Goats, Poultry and Rabbits, and Relative to Other Matters Deemed Useful in the Conducting of the Affairs of a Labourer's Family.* London: C. Clement, 1822.

Cohen, Marjorie Griffin. *Women's Work, Markets and Economic Development in Nineteenth Century Ontario.* Toronto, ON: University of Toronto Press, 1988.

Collier, Mary. "The Woman's Labour: An Epistle to Mr Stephen Duck in Answer to His Late Poem, Called 'The Thresher's Labour.'" London: J. Roberts, 1739. Accessed 3 December 2019. https://www.usask.ca/english/barbauld/related_texts/collier.html.

Colpitts, George. "Food Energy and the Expansion of the Canadian Fur Trade." In *Powering Up Canada: A History of Power, Fuel and Energy from 1600*, edited by R.W. Sandwell, 39–58. Montreal and Kingston: McGill-Queen's University Press, 2016.

Connell, Raewyn. "Foreword: Masculinities in the Sociocene." In "Men and Nature: Hegemonic Masculinities and Environmental Change," edited by Sherilyn MacGregor and Nicole Seymour. RCC *Perspectives* 4 (2017).

Connolly, Linda. *The Irish Women's Movement: From Revolution to Devolution.* Hampshire: Palgrave Macmillan, 2002.

Connolly, Linda, and Tina O'Toole. *Documenting Irish Feminism: The Second Wave.* Dublin: Woodfield Press, 2005.

Conway, Moncure D. *Travels in Kensington: With Notes on Decorative Arts and Architecture in England.* London: Trubner, 1882.

Cooper, Gail. *Air-Conditioning America: Engineers and the Controlled Environment, 1900–1960.* Baltimore, MD: Johns Hopkins Press, 1998.

Cooper, Nicholas. *The Opulent Eye: Late Victorian and Edwardian Taste in Interior Design.* New York: Watson-Guptill, 1977.

Cooper, Tim. "Why We Still Need a Human History in the Anthropocene," blog post, 6 February 2014. Accessed 8 July 2020. https://blogs.exeter.ac.uk/historyenvironmentfuture/2014/02/06/167.

Corrigan, Vawn. *Irish Aran: History, Tradition, Fashion*. Dublin: O'Brien Press, 2019.

Cottrell, Fred. *Energy and Society: The Relation between Energy, Social Change, and Economic Development*. Westport, CT: Greenwood Press, 1955.

Cowman, Krista. "'From the Housewife's Point of View': Female Citizenship and the Gendered Domestic Interior in Post-First World War Britain, 1918–1928." *English Historical Review* 130, no. 543 (2015): 352–83.

Crawford, Elizabeth. *Enterprising Women: The Garretts and Their Circle*. London: Francis Boutle Publishers, 2002.

– "Spirited Women of Gower Street: The Garretts and Their Circle." Unpublished conference paper, UCL Bloomsbury Project, 2011. www.ucl. ac.uk/bloomsbury-project/articles/events/conference2011/crawford.pdf.

Cronon, William. *Nature's Metropolis: Chicago and the Great West*. New York: W.W. Norton, 1991.

Crosby, Alfred W. *Children of the Sun: A History of Humanities Unappeasable Appetite for Energy*. New York: W.W. Norton, 2006.

Crutzen, Paul J., and Eugene F. Stoermer. "The 'Anthropocene.'" *IGBP Newsletter* (International Geosphere-Biosphere Programme) 41 (2000): 17–18.

Daggett, Cara New. *The Birth of Energy: Fossil Fuels, Thermodynamics, and the Politics of Work*. Durham and London: Duke University Press, 2019.

d'Allemagne, Henry-René. "Histoire du luminaire depuis l'époque romaine jusqu'au XIXe siècle, par Henry-René d'Allemagne." *Bibliothèque de l'école des chartes* 52, no. 1 (1891): 475–6. Accessed 19 September 2019. https:// www.e-rara.ch/zut/content/pageview/4278686.

Daly, Mary E. *Sixties Ireland: Reshaping the Economy, State and Society 1957–1973*. Cambridge, UK: Cambridge University Press, 2016.

Davidoff, Leonore, and Catherine Hall. *Family Fortunes: Men and Women of the English Miiddle Class, 1750–1850*. Chicago, IL: University of Chicago Press, 1987.

Davidson, Caroline. *A Woman's Work Is Never Done: A History of Housework in the British Isles, 1650–1950*. London: Chatto and Windus, 1982.

Davin, Anna. "Imperialism and Motherhood." *History Workshop Journal* 5 (spring 1978): 9–66.

Davis, Lance, Robert Gallman, and Teresa Hutchins. "The Decline of US Whaling: Was the Stock of Whales Running Out?" *Business History Review* 62, no. 4 (1988): 569–95.

Debeir, Jean-Claude, Jean-Paul Deléage, and Daniel Hemery. *In the Servitude of Power: Energy and Civilization through the Ages*. Translated by John Barzman. Atlantic Highlands, NJ: Zed Books, 1991.

de Jong, Jutta. "Bergarbeiterfrauen – oder Die andere Arbeit für den Bergbau." In *Frauen und Bergbau: Zeugnisse aus 5 Jahrhunderten. Ausstellung des Deutschen Bergbau-Museums Bochum vom 29. August bis 10. Dezember 1989*, edited by Evelyn Kroker, 70–5. Bochum: Deutsches Bergbau-Museum, 1989.

– ed. *Kinder, Küche, Kohle – und viel mehr! Bergarbeiterfrauen aus drei Generationen erinnern sich.* Essen, Germany: Klartext, 1991.

– "'Sklavin' oder 'Hausdrache'? Frauen in Bergarbeiterfamilien." In *Eine Partei in ihrer Region: Zur Geschichte der SPD im Westlichen Westfalen*, edited by Bernd Faulenbach and Günther Högl, 45–50. Essen, Germany: Klartext, 1988.

– ed. *"… und die Wäsche, die war schwarz, ja, wie die Kohle!" Erzählungen von der Großen Wäsche der Bergarbeiterfrauen, zusammengetragen vom "Gesprächskreis Lebenserfahrung von Frauen in Bergarbeiterfamilien."* 2nd ed. Herten: n.p., 1988.

Department for Business, Energy and Industrial Strategy. *2016 UK Greenhouse Gas Emissions, Final Figures.* London: Department for Business, Energy and Industrial Strategy, 6 February 2018. Accessed 24 June 2019. https://assets.publishing.service.gov.uk/government/uploads/system/uploads/attachment_data/file/680473/2016_Final_Emissions_statistics.pdf.

Department for Communities and Local Government. *The English Indices of Deprivation 2010.* London: Department for Communities and Local Government, 2011. Accessed 10 November 2019. https://www.gov.uk/government/statistics/english-indices-of-deprivation-2010.

de Vries, Jan. *The Industrious Revolution: Consumer Behavior and the Household Economy, 1650 to the Present.* Cambridge, UK: Cambridge University Press, 2008.

Dolata-Kreutzkamp, Petra. *Die deutsche Kohlenkrise im nationalen und transatlantischen Kontext.* Wiesbaden: VS Verlag für Sozialwissenschaften, 2006.

Doyle, Clodagh. "A Study of Traditional Hearth Furniture and the Hearth in Irish Rural Society, 1750–1950." MA thesis, National University of Ireland, Cork, Ireland, 1999.

Dresser, Christopher. *Studies in Design.* London: Cassell, Petter, and Calpin, 1879.

Duck, Stephen. *The Thresher's Labour.* London: Merlin Press, (1730) 1989.

Eckirch, Roger A. *At Day's Close: Night in Times Past.* New York: W.W. Norton, 2005.

Edis, Robert W. "Internal Decoration." In *Our Homes and How to Make Them Healthy*, edited by Shirley F. Murphy. London: Cassell and Co., 1883.

Edwards, Clive. "Establishing Stability: Conforming to Type in British House Furnishings, 1860–1910." In *The Objects and Textures of Everyday Life in Imperial Britain*, edited by Deirdre H. McMahon and Janet C. Myers. Farnham, UK: Ashgate, 2016.

Edwards, Jason, and Imogen Hart, eds. *Rethinking the Interior, c. 1867–1896*. Farnham, UK: Ashgate, 2010.

Ehrenreich, Barbara and Deidre English. *For Her Own Good: 150 Years of the Experts' Advice to Women*. New York: Anchor/Dougleday, 1978.

Einheit: Organ der Industriegewerkschaft Bergbau und Energie 23, no. 2 (1970).

Electricity Supply Board. "Electricity Supply Board Forty Third Annual Report for Year Ended 31st March 1970." Dublin: Electricity Supply Board, 1970.

– "Save Time and Labour: Bring Water to Your Home the Electric Way." Dublin: Electricity Supply Board, n.d.

– "What a Unit Can Do." Dublin: Electricity Supply Board, 1960.

Enquire Within upon Everything, 89th rev. ed. London: Houlston and Sons, 1894.

Etzler, John Adolphus. *The Paradise within the Reach of All Men, without Labour, by Powers of Nature and Machinery: An Address to All Intelligent Men*. Pittsburgh, PA, 1833.

Falkus, M.E. "The Early Development of the British Gas Industry, 1790–1815." *Economic History Review* 35, no. 2 (1982): 217–34.

Ferriter, Diarmuid. *The Transformation of Modern Ireland 1900–2000*. London: Profile Books, 2004.

Ferry, Emma. "'Any Lady Can Do This without Much Trouble…': Class and Gender in *The Dining Room* (1878)." *Interiors* 5, no. 2 (2014): 141–59.

– "Decorators May Be Compared to Doctors: An Analysis of Rhoda and Agnes Garrett's *Suggestions for House Decoration in Painting, Woodwork and Furniture* (1876)." *Journal of Design History* 16, no. 1 (2003): 15–33.

– "Information for the Ignorant and Aid for the Advancing: Macmillan's 'Art at Home' Series, 1876–1883." In *Design and the Modern Magazine*, edited by Jeremy Aynsley and Kate Forde, 134–55. Manchester, UK: Manchester University Press, 2007.

– "'The Other Miss Faulkner': Mrs Orrinsmith and the 'Art at Home' Series," *Journal of William Morris Studies* 23 (summer 2011): 47–64.

Fitzpatrick, Mona. "An Enemy of Drudgery." *REO News*, 1968, 8–9.

Fleetwood, David. "The Electrification of Scotland." In *Transforming the Countryside: The Electrification of Rural Britain*, edited by Paul Brassley, Jeremy Burchardt, and Karen Sayer, 69–82. London: Routledge, 2017.

Fletcher, Roger. "History from Below Comes to Germany: The New History Movement in the Federal Republic of Germany." *Journal of Modern History* 60, no. 3 (1988): 557–68.

Floud, Roderick, Robert William Fogel, Bernard Harris, and Sok Chul Hong. *The Changing Body: Health, Nutrition and Human Development in the Western World Since 1700*. Cambridge, UK: Cambridge University Press 2011.

Folbre, Nancy. "Hearts and Spades: Paradigms of Household Economics." *World Development* 14, no. 2 (1986): 245–55.

– "The Unproductive Housewife: Her Evolution in Nineteenth-Century Economic Thought." *Signs: Journal of Women in Culture and Society* 16, no. 31 (1991): 463–84.

Forty, Adrian. *Objects of Desire: Design and Society since 1750*. London: Thames and Hudson, 1986.

Fouquet, Roger. *Heat, Power and Light: Revolutions in Energy Services*. Cheltenham, UK, and Northampton, MA: Edward Elgar Publishing, 2008.

– "The Slow Search for Solutions: Lessons from Historical Energy Transitions by Sector and Service." *Energy Policy* 38 (2010): 6,586–96.

Fouquet, Roger, and S.N. Broadberry. "Seven Centuries of European Economic Growth and Decline." *Journal of Economic Perspectives* 29, no. 4 (2015): 227–44.

Fouquet, Roger, and Peter Pearson. "Seven Centuries of Energy Services: The Price and Use of Light in the United Kingdom (1300–2000)." *Energy Journal* 27, no. 1 (2006): 139–77.

Fraser, Nancy. "Behind Marx's Hidden Abode: For an Expanded Conception of Capitalism." *New Left Review* 86 (March–April 2014): 55–72.

French, Daniel. *When They Hid the Fire: A History of Electricity and Invisible Energy in America*. Pittsburgh, PA: University of Pittsburgh Press, 2017.

Friedan, Betty. *The Feminine Mystique*. New York: Norton, 1963.

Friedmann, Harriet. "World Market, State, and Family Farm: Social Bases of Household Production in the Era of Wage Labour." *Comparative Studies in Society and History* 20 (1978): 545–86.

Garrett, Rhoda, and Agnes Garrett. *Suggestions for House Decoration in Painting, Woodwork and Furniture*. London: Macmillan, 1877.

Garvin, Tom. *Preventing the Future: Why Was Ireland So Poor for So Long?* Dublin: Gill and McMillan, 2004.

Ghosh, Amitav. *The Great Derangement*, Chicago, IL: University of Chicago Press, 2016.

Gibson, Chris, Lesley Head, Nick Gill, and Gordon Waitt. "Climate Change and Household Dynamics: Beyond Consumption, Unbounding Sustainability." *Transactions of the Institute of British Geographers* 36, no. 1 (2011): 3–8.

Gillott, W.A. "Domestic Load Building: A Few Suggestions upon Propaganda Work." *Journal of the Institution of Electrical Engineers* 61 no. 315 (1923): 197–202.

"Glasgow International Exhibition, Industrial Art, Women's Arts and Industries," special edition of *Art Journal*, December 1888.

Gleeson, Colin, and Patrick Logue. "Referendum on 'Sexist' Reference to Women's Place in the Home Postponed." *Irish Times*, 5 September 2018.

Goch, Stefan. "Betterment without Airs: Social, Cultural, and Political Consequences of De-industrialization in the Ruhr." *International Review of Social History* 47, no. 10 (2002): 87–111.

– *Eine Region im Kampf mit dem Strukturwandel: Bewälligung von Strukturwandel und Strukturpolitik im Ruhrgebiet*. Essen, Germany: Klartext, 2002.

Goldstein, Carolyn. *Creating Consumers: Home Economists in Twentieth-Century America*. Chapel Hill, NC: University of North Carolina Press, 2012.

González de Molina, Manuel, and Victor M. Toledo. *The Social Metabolism: A Socio-Ecological Theory of Historical Change*. London: Springer, 2014.

Gooday, Graeme. *Domesticating Electricity: Technology, Uncertainty and Gender, 1880–1914*. London: Pickering and Chatto, 2008.

Gooday, Graeme, and Abigail Harrison Moore. "True Ornament?: The Art and Industry of Electric Lighting in the Home, 1889–1902." In *Art Versus Industry? New Perspectives on Visual and Industrial Cultures in Nineteenth-Century Britain*, edited by Kate Nichols, Rebecca Wade, and Gabriel Williams, 158–78. Manchester, UK: Manchester University Press, 2015.

Gordon, Mrs J.E.H. [Alice]. *Decorative Electricity, with a chapter on Fire Risks by J.E.H. Gordon*. London: Sampson Low, et al., 1891.

Gray, Jane. "Rural Industry and Uneven Development: The Significance of Gender in the Irish Linen Industry." *Journal of Peasant Studies* 20 (1993): 590–611.

Gray, Robert. "Factory Legislation and the Gendering of Jobs in the North of England." *Gender and History* 5 (1993): 56–80.

Greenberg, Dolores. "Energy Flow in a Changing Economy, 1815–1880." In *An Emerging Independent American Economy, 1815–1875*, edited by Joseph R. Frese and Jacob Judd, 29–58. Tarrytown, NY: Sleepy Hollow Restorations, 1980.

– "Energy, Power, and Perceptions of Social Change in the Early Nineteenth Century." *American Historical Review* 95 (1990): 693–714.

– "Reassessing the Power Patterns of the Industrial Revolution: An Anglo-American Comparison." *American Historical Review* 87 (1982): 1237–61.

Greer, Carlotta. *Foods and Homemaking*. Boston, MA: Allyn and Bacon, 1937.

Griffin, Emma. *Bread Winner: An Intimate History of the Victorian Economy*. New Haven, CT: Yale University Press, 2020.

– *Liberty's Dawn: A People's History of the Industrial Revolution*. New Haven, CT: Yale University Press, 2013.

Gullickson, Gay. *The Spinners and Weavers of Auffay: Rural Industry and the Sexual Division of Labour in a French Village, 1750–1850*. Cambridge, UK: Cambridge University Press, 1986.

Günter, Janne. *Mündlicher Geschichtsschreibung: Alte Menschen im Ruhrgebiet erzählen erlebte Geschichte*. Mühlheim/Ruhr: Westarp Verlag, 1982.

Hagemann, Gro, and Hege Roll-Hansen, eds. *Twentieth-Century Housewives: Meanings and Implications of Unpaid Work*. Oslo: Oslo Academic Press, 2005.

Haley, Bruce. *The Healthy Body and Victorian Culture*. Cambridge, MA: Harvard University Press, (1978) 2014.

Hall, Catherine. *White, Male and Middle Class: Explorations in Feminism and History*. Cambridge, UK: Polity, 1992.

Hall, Valerie G. *Women at Work, 1860–1939: How Different Industries Shaped Women's Experiences*. Woodbridge, UK: Boydell Press, 2013.

Hammond, Robert. Hammond Electric Light and Power Supply Co. *The Electric Light in Our Homes*. London: Frederick Warne and Co., 1884.

Hannah, Leslie. *Electricity before Nationalisation: A Study of the Development of the Electricity Supply Industry in Britain to 1948*. London: Macmillan, 1979.

– *Engineers, Managers, and Politicians: The First Fifteen Years of Nationalised Electricity Supply in Britain*. London: Macmillan, 1982.

Hannam, June. "Women as Paid Organizers and Propagandists for the British Labour Party between the Wars." *International Labor and Working-Class History* 77 (spring 2010): 69–88.

Haraway, Donna. "Anthropocene, Capitalocene, Plantationocene, Chthulucene: Making Kin." *Environmental Humanities* 6, no. 1 (2015): 159–65.

– "SF: Science Fiction, Speculative Fabulation, String Figures, So Far." *Ada: A Journal of Gender, New Media and Technology* 11, no. 3 (2013).

Harris, Carmen V. "The South Carolina Home in Black and White: Race, Gender and Power in Home Demonstration Work." *Agricultural History* 93, no. 3 (summer 2019): 477–501.

Harris, Howell John. "Inventing the US Stove Industry, c. 1815–1875: Making and Selling the First Universal Consumer Durable." *Business History Review* 82, no. 4 (winter 2008): 701–33.

– "Conquering Winter: US Consumers and the Cast-Iron Stove." In *Comfort in a Lower Carbon Society*, special edition of *Building Research and Information*, edited by Elizabeth Shove, Heather Chappells, and Loren Lutzenhiser, 33–46. New York: Routledge, 2010.

Harrison Moore, Abigail. "Designing Energy Use in a Rural Setting: A Case Study of Philip Webb at Standen." In "Off-Grid Empire: Rural Energy Consumption in Britain and the British Empire, 1850–1960," edited by Abigail Harrison Moore and R.W. Sandwell, special issue of *History of Retailing and Consumption* 4, no. 1 (2018): 28–42.

Harrison Moore, Abigail, and Graeme Gooday. "Decorative Electricity: Standen and the Aesthetics of New Lighting Technologies in the Nineteenth Century Home." *Nineteenth-Century Contexts: An Interdisciplinary Journal* 35, no. 4 (2013): 363–83.

Harrison Moore, Abigail, and R.W. Sandwell, eds. "Off-Grid Empire: Rural Energy Consumption in Britain and the British Empire, 1850–1960." Special issue of *History of Retailing and Consumption* 4, no. 1 (2018).

Hartman, Mary S. *The Household and the Making of History: A Subversive View of the Western Past*. Cambridge, UK: Cambridge University Press, 2004.

Haslett, Caroline. "Electricity in the Home." *Journal of the Royal Society of Arts* 95, no. 4749 (1947): 639–54.

Haslett, Caroline, ed. *Electrical Handbook for Women*. London: Hodder and Stoughton/English Universities Press, 1934.

Haslett, Caroline, and Elsie Emitt Edwards. *Teach Yourself Household Electricity*. London: English Universities Press, 1939.

Haslett, Caroline, et al. *Ninth Annual Report of the Electrical Association for Women*. London: Electrical Association for Women, 1934.

Haweis, Mrs [Mary Eliza]. *The Art of Decoration*. London: Chatto and Windus, 1881.

Hayden, Dolores. *The Grand Domestic Revolution: A History of Feminist Designs for American Homes, Neighborhoods and Cities*. Cambridge, MA: MIT Press, 1982.

Heald, Henrietta. *Magnificent Women and Their Revolutionary Machines*. London: Unbound, 2019.

Hecht, Gabrielle. *Being Nuclear: Africans and the Global Uranium Trade*. Cambridge, MA: MIT Press, 2012.

Hewison, Richard. *The Heritage Industry: Britain in a Climate of Decline*. London: Methuen Publishing, 1987.

Hewitt, Margaret. *Wives and Mothers in Victorian Industry*. London: Rockliff, 1958.

Hilton, Matthew. *Consumerism in Twentieth-Century Britain: The Search for a Historical Movement*. Cambridge, UK: Cambridge University Press 2003.

Holmes, K.B. "Making Masculinity: Land, Body, Image in Australia's Mallee Country." In "Visions of Australia: Environments in History," edited by Christof Mauch, Ruth Morgan, and Emily O'Gorman, special edition of *RCC Perspectives: Transformations in Environment and Society* 2 (2017): 39–48.

Horowitz, Roger, and Arwen Mohun. *His and Hers: Gender, Technology and Consumption*. Charlottesville, VA: University of Virginia Press, 1998.

Horrell, Sara, and Jane Humphries. "The Origins and Expansion of the Male Breadwinner Family: The Case of Nineteenth-Century Britain." *International Review of Social History* 42, supplement (1997): 25–64.

Houston, Mary. "The Hon. Lady Parsons (Hon. Fellow)." *Transactions of the North East Coast Institution of Engineers and Shipbuilders* 50 (1933–1934): 181–2.

Huber, Matthew T. *Lifeblood: Oil, Freedom and the Forces of Capital*. Minneapolis, MN: University of Minnesota Press, 2013.

Hudson, Pat and W. Robert Lee, eds. *Women's Work and the Family Economy in Historical Perspective*. Manchester, UK: Manchester University Press, 1990.

Huebel, Sebastian. *Fighter, Worker, and Family Man: German-Jewish Men and Their Gendered Experiences in Nazi Germany, 1933–1941*. Toronto, ON: University of Toronto Press, 2021.

Hughes, Thomas P. *Networks of Power: Electrification in Western Society, 1880–1930* Baltimore, MD: Johns Hopkins University Press, 1993.

Humphries, Jane. "Class Struggle and the Persistence of the Working Class Family." *Cambridge Journal of Economics* 1 (1977): 241–58.

Hunt, Alex. "Anxiety and Social Explanation: Some Anxieties about Anxiety." *Journal of Social History* 32, no. 3 (spring 1999): 509–28.

Iacovetta, Franca, Valerie J. Korinek, and Marlene Epp, eds. *Edible Histories, Cultural Politics: Towards a Canadian Food History*. Toronto, ON: University of Toronto Press, 2012.

Ihde, Don. *Bodies in Technology*. Minneapolis, MN: University of Minnesota Press, 2002.

Illich, Ivan. *Shadow Work*. London: Marion Boyers, 1981.

Isenstadt, Sandy. *Electric Light: An Architectural History*. Cambridge, MA: MIT Press, 2018.

Jellison, Katherine. *Entitled to Power: Farm Women and Technology, 1913–63*. Chapel Hill, NC: University of North Carolina Press, 1993.

Johnson, Bob. *Carbon Nation: Fossil Fuels in the Making of Modern Culture*. Lawrence, KS: University of Kansas Press, 2014.

Johnston, Thomas. *Memories*. London: Collins, 1952.

Jones, Christopher F. "The Carbon-Consuming Home: Residential Markets and Energy Transitions." *Enterprise and Society* 12, no. 4 (2011): 790–823.

– "A Landscape of Energy Abundance: Anthracite Coal Canals and the Roots of American Fossil Fuel Dependence, 1820–1860." *Environmental History* 15, no. 3 (2010): 449–84.

– *Routes of Power: Energy and Modern America*. Cambridge, MA: Harvard University Press, 2014.

Kander, Astrid, Paolo Malanima, and Paul Warde. *Power to the People: Energy in Europe over the Last Five Centuries*. Princeton, NJ: Princeton University Press, 2014.

Kaschuba, Wolfgang. "Volkskultur und Arbeiterkultur als symbolische Ordnungen: Einige volkskundliche Anmerkungen zur Debatte um Alltags- und Kulturgeschichte." In *Alltagsgeschichte: Zur Rekonstruktion historischer Erfahrungen und Lebensweisen*, edited by Alf Lüdtke, 191–223. Frankfurt: Campus Verlag, 1989.

Kearney, Joe. "Before the Light." In *Then There Was Light: Stories Powered by the Rural Electrification Scheme in Ireland*, edited by Joe Kearney and P.J. Cunningham, 53–5. Bray, Ireland: Ballpoint Press, 2016.

Kelley, Victoria. "Housekeeping: Shine, Polish, Gloss and Glaze as Surface Strategies in the Domestic Interior." In *The Objects and Textures of Everyday Life in Imperial Britain*, edited by Deirdre H. McMahon and Janet C. Myers, 93–111. Farnham, UK: Ashgate, 2016.

Kiechle, Melanie. "Navigating by Nose: Fresh Air, Stench Nuisance and the Urban Environment, 1840–1880." *Journal of Urban History* 42, no. 4 (2016): 753–71.

– *Smell Detectives: An Olfactory History of Nineteenth-Century Urban America.*
Seattle, WA: University of Washington Press, 2017.

Kline, Ronald. "Agents of Modernity: Home Economics and Rural
Electrification, 1925–1950." In *Rethinking Home Economics: Women and the
History of a Profession*, edited by Sarah Stage and Virginia Vincenti, 237–52.
Ithaca, NY: Cornell University Press, 1997.

Kriedte, Peter, Hans Medick, and Jurgen Schlumbohm. *Industrialization before
Industrialization: Rural Industry in the Genesis of Capitalism.* Translated by
Beate Schempp. Cambridge, UK: Cambridge University Press, 1981.

Landes, Joan B. *Women and the Public Sphere in the Age of the French
Revolution.* Ithaca, NY: Cornell University Press, 1988.

Langland, Elizabeth. *Nobody's Angels, Middle-Class Women and Domestic
Ideology in Victorian Culture.* Ithaca, NY: Cornell University Press, 1995.

Larkin, Emmet. "The Devotional Revolution in Ireland, 1850–75." *American
Historical Review* 77, no. 3 (1972): 625–52.

Lasch, Christopher. *Haven in a Heartless World: The Family Besieged.* New
York: Basic Books, 1977.

Laslett, Peter. *The World We Have Lost: England Before the Industrial Age.*
London: Methuen, 1965.

Leach, Melissa, and Cathy Green. "Gender and Environmental History: From
Representation of Women and Nature to Gender Analysis of Ecology and
Politics." *Environment and History* 3, no. 3 (1997): 343–70.

LeCain, Timothy. *The Matter of History: How Things Create the Past.*
Cambridge, UK: Cambridge University Press, 2017.

Lehning, James Robert. *Peasant and French: Cultural Contact in Rural France
During the Nineteenth Century.* Cambridge, UK: Cambridge University
Press, 1995.

– *The Peasants of Marlhes: Economic Development and Family Organization
in Nineteenth Century France.* Chapel Hill: University of North Carolina
Press, 1980.

Lerner, Gerda. *Living with History/Making Social Change.* Chapel Hill, NC:
University of North Carolina Press, 2009.

Levine, David. *Reproducing Families: The Political Economy of English
Population History.* Cambridge, UK: Cambridge University Press, 1987.

Lewis, Jane, ed. *Labour and Love: Women's Experience of Home and Family.*
Cambridge, UK: Blackwells, 1986.

Liddington, Jill. *Rebel Girls: How Votes for Women Changed Edwardian Lives.*
London: Virago/Little, Brown Book Group, 2006.

Lindenberger, Thomas, Michael Wildt, Lyndal Roper, and Martin Chalmers.
"Radical Plurality: History Workshops as a Practical Critique of
Knowledge." *History Workshop* 33 (1992): 73–99.

Lindsay, R. Bruce. *Energy: Historical Development of a Concept.*
  Stroudsburg, PA: Dowden, Hutchinson, and Ross, 1975.
Llewellyn, Mark. "Designed by Women and Designing Women: Gender,
  Planning and the Geographies of the Kitchen in Britain, 1917–1946."
  Cultural Geographies 11, no. 1 (2004): 42–60.
Locker, Anne. "Partridge, Margaret Mary (1891–1967)." *Oxford
  Dictionary of National Biography*, 2018. https://doi.org/10.1093/
  odnb/9780198614128.013.110230.
Long, Jane. *Conversations in Cold Rooms: Women, Work and Poverty in
  Nineteenth-Century Northumberland.* London: Royal Historical Society, 1999.
Luber, Steven, and W.D. Kingery. *History from Things.* Washington, DC:
  Smithsonian Press, 1993.
Luckin, Bill. *Questions of Power: Electricity and Environment in Inter-war
  Britain.* Manchester, UK: Manchester University Press, 1990.
– "Revisiting the Idea of Degeneration in Urban Britain, 1830–1900." *Urban
  History* 33, no. 2 (2006): 234–52.
Lüdtke, Alf. "Einleitung: Was ist und wer treibt Alltagsgeschichte?"
  In *Alltagsgeschichte: Zur Rekonstruktion historischer Erfahrungen und
  Lebensweisen*, edited by Alf Lüdtke, 9–47. Frankfurt: Campus Verlag, 1989.
Luxton, Meg. *More than a Labour of Love: Three Generations of Women's Work
  in the Home.* Toronto, ON: Women's Educational Press, 1980.
MacDonald, Eva M. "How the Cooking Stove Transformed the Kitchen in Pre-
  Confederation Ontario." *Culinary Chronicles*, no. 43 (winter 2005): 3–12.
MacGregor, Sherilyn. "Go Ask 'Gladys': Why Gender Matters in Energy
  Consumption Policy and Research." LSE Blog, 9 February 2016. Accessed
  8 July 2020. https://blogs.lse.ac.uk/politicsandpolicy/go-ask-gladys-why-
  gender-matters-in-energy-consumption-policy-and-research.
MacPherson, David. "Women, Home and Irish Identity: Discourses of
  Domesticity in Ireland, C. 1890–1922." PhD thesis, Birkbeck College, 2004.
Malm, Andreas. *Fossil Capital: The Rise of Steam Power and the Roots of Global
  Warming.* London: Verso Press, 2016.
– *The Progress of This Storm: Nature and Society in a Warming World.* London:
  Verso Press, 2018.
Malm, Andreas, and Alf Hornborg. "The Geology of Mankind? A Critique of
  the Anthropocene Narrative." *Anthropocene Review* 1, no. 1 (2014): 62–9.
Manning, Maurice, and Moore McDowell. *Electricity Supply in Ireland: The
  History of the ESB.* Dublin: Gill and Macmillan, 1984.
Marshall, Dorothy. *The English Domestic Servant in History.* London: Historical
  Association, 1st ed., (1949) 1969.
Martiny, Martin, and Hans-Jürgen Schneider, eds. *Deutsche Energiepolitik seit
  1945 – Vorrang für die Kohle: Dokument und Materialien zur Energiepolitik
  der Industriegewerkschaft Bergbau und Energie.* Cologne: Bund-Verlag, 1981.
Marx, Karl. *Capital, Vol. III.* New York: Vintage, 1981.

Massell, David. *Quebec Hydropolitics: The Peribonka Concessions of the Second World War*. Montreal and Kingston: McGill-Queen's University Press, 2011.

Mathias, H.S. "Leaves from an Inspector's Note Book." Read before the third congress of the Canadian Public Health Association, Regina, Saskatchewan. *Public Health Journal* 4, no. 10 (1913): 563.

McAfee, Kathleen. "The Politics of Nature in the Anthropocene." In "Whose Anthropocene: Revisiting Chakrabarty's 'Four Theses,'" edited by Robert Emmett and Thomas Lekan, special edition of *RCC Perspectives: Transformations in Environment and Society* 2 (2016): 65–72.

McBride, T.M. *The Domestic Revolution: The Modernisation of Household Service in England and France, 1820–1920*. London: Croom Helm, 1976.

McCarty, Elizabeth A. "Irene May Lovelock (1896–1974)." In *Dictionary of National Biography*. Oxford, UK: Oxford University Press, 2004.

McCoole, Sinead. *No Ordinary Women: Irish Female Activists in the Revolutionary Years 1900–1923*. Dublin: O'Brien Press, 2015.

McGraw, Judith A. "No Passive Victims, No Separate Spheres: A Feminist Perspective on Technology's History." In *In Context: History and the History of Technology: Essays in Honor of Melvin Kranzberg*, edited by Stephen Cutcliffe and Robert C. Post, 172–91. Bethlehem, PA: Lehigh University Press, 1989.

McNeill, J.R. "Cheap Energy and Ecological Teleconnections of the Industrial Revolution, 1780–1920." Forum: Environmental History of Energy Transitions, *Environmental History* 24, no. 3 (2019): 30–3.

– *Something New under the Sun: An Environmental History of the Twentieth Century World*. New York: Norton, 2000.

McNeill, J.R., and Peter Engelke. *The Great Acceleration: An Environmental History of the Anthropocene since 1945*. Cambridge, MA: Belknap Press of Harvard University Press, 2014.

Medick, Hans. "'Missionare im Ruderboot'?: Ethnologische Erkenntnisweisen als Herausforderung an die Sozialgeschichte." In *Alltagsgeschichte: Zur Rekonstruktion historischer Erfahrungen und Lebensweisen*, edited by Alf Lüdtke, 48–84. Frankfurt: Campus Verlag, 1989.

– "The Proto-Industrial Family Economy: The Structural Function of the Household and Family During the Transition from Peasant Society to Industrial Capitalism." In *Essays in Social History* 2, edited by Pat Thane and Anthony Sutcliffe. Oxford: Clarendon Press, 1986.

– "Structures and Functions of Population Development." In *Industrialization before Industrialization: Rural Industry in the Genesis of Capitalism*, edited by Peter Kriedte, Hans Medick, Jurgen Schlumbohm, 74–93. Translated by Beate Schempp. Cambridge, UK: Cambridge University Press, 1981.

Meehan, Ciara. "Modern Wife, Modern Life." Dublin: National Print Museum, 2015.

Meller, Helen. "Women and Citizenship: Gender and the built environment in British Cities, 1870–1930." In *Cities of Ideas: Governance and Citizenship in Urban Britain*, edited by Robert Colls and Richard Rodger, 234–5. Aldershot, UK: Ashgate, 2004.

Melosi, Martin. *Coping with Abundance: Energy and Environment in Industrial America*. New York: Alfred Knopf, 1985.

– *The Sanitary City: Environmental Services in Urban America from Colonial Times to the Present*. Baltimore, MD: Johns Hopkins University Press, 2000.

Merchant, Carolyn. *The Death of Nature; Ecological Revolutions: Nature, Gender and Science in New England*. San Francisco, CA: Harper and Row, 1980.

– *Earthcare: Women and the Environment*. New York: Routledge, 1995.

Messenger, Rosalind. *The Doors of Opportunity: A Biography of Dame Caroline Haslett*. London: Femina Books, 1967.

Meyer, John M. "Politics In – but Not Of – the Anthropocene." In "Whose Anthropocene? Revisiting Chakrabarty's 'Four Theses,'" edited by Robert Emmett and Thomas Lekan, special edition of RCC *Perspectives: Transformations in Environment and Society* 2 (2016): 47–51.

Meyer-Renschhausen, Martin. *Energiepolitik in der BRD von 1950 bis heute*. Cologne: Pahl-Rugenstein, 1977.

Miller, Ian Jared, and Paul Warde. "Energy Transitions as Environmental Events." Forum: Environmental History of Energy Transitions, *Environmental History* 24, no. 2 (2019): 464–71.

Mitchell, Timothy. *Carbon Democracy: Political Power in the Age of Oil*. London and New York: Verso, 2011.

Moir, Margaret, and Caroline Haslett, "Preface," in *Electrical Handbook for Women*, edited by Caroline Haslett. London: Hodder and Stoughton/ English Universities Press, 1934.

Möllers, Nina, and Karin Zachmann, eds. *Past and Present Energy Societies: How Energy Connects Politics*. Bielefeld, Germany: Transcript Verlag, 2012.

Montrie, Chad. *Making a Living: Work and Environment in the United States*. Chapel Hill, NC: University of North Carolina Press, 2008.

Moore, Jason W. "The End of Cheap Nature, or: How I Learned to Stop Worrying about 'the' Environment and Love the Crisis of Capitalism." In *Structures of the World Political Economy and the Future of Global Conflict and Cooperation*, edited by Christian Suter and Christopher Chase-Dunn, 1–31. Berlin/Zurich: LIT, 2014.

Moore, Jason W., ed. *Anthropocene or Capitalocene? Nature, History, and the Crisis of Capitalism*. Oakland, CA: PM Press, 2016.

Mosley, Stephen. *The Chimney of the World: A History of Smoke Pollution in Victorian and Edwardian Manchester*. London and New York: Routledge, 2001.

– "Clearing the Air: Can the 1956 Clean Air Act Inform New Legislation?" History and Policy, 24 July 2017. Accessed 24 November 2019. http://www.historyandpolicy.org/policy-papers/papers/clearing-the-air-can-the-1956-clean-air-act-inform-new-legislation.

Mrs Gaskell. *Cranford*. London: Macmillan, (1853) 1923.

Murphy, Michelle. *Sick Building Syndrome and the Problem of Uncertainty: Environmental Politics, Technoscience and Women Workers*. Durham and London: Duke University Press, 2006.

Nash, Linda. *Inescapable Ecologies: A History of Environment, Disease and Knowledge*. Berkeley, CA: University of California Press, 2006.

Nead, Lynda. *Victorian Babylon: People, Streets and Images in Nineteenth-Century London*. New Haven, CT: Yale University Press, 2000.

Neimanis, Astrida, Cecilia Asberg, and Johan Hedren. "Four Problems, Four Direction for Environmental Humanities: Toward Critical Post-humanities for the Anthropocene." *Ethics and the Environment* 20, no. 1 (2014): 67–142.

Newton, Janice. *The Feminist Challenge to the Canadian Left, 1900–1918*. Montreal and Kingston: McGill-Queen's University Press, 1995.

Nicholson, Linda. *Gender and History: The Limits of Social Theory in the Age of the Family*. New York: Columbia University Press, 1986.

Niethammer, Lutz, ed. *"Die Jahre weiß man nicht, wo man die heute hinsetzen soll": Faschismuserfahrungen im Ruhrgebiet*. Bonn, Germany: Dietz, 1983.

– *"Hinterher merkt man, daß es richtig war, daß es schiefgegangen ist": Nachkriegserfahrungen im Ruhrgebiet*. Bonn, Germany: Dietz, 1983.

Niethammer, Lutz, and Alexander von Plato, eds. *"Wir kriegen jetzt andere Zeiten": Auf der Suche nach der Erfahrung des Volkes in nachfaschistischen Ländern*. Bonn, Germany: Dietz, 1985.

Nonn, Christoph. *Die Ruhrbergbaukrise: Entindustrialisierung und Politik, 1958–1969*. Göttingen, Germany: Vandenhoeck and Ruprecht, 2001.

– "Vom Naturschutz zum Umweltschutz: Luftreinhaltung in Nordrhein-Westfalen zwischen fünfziger und frühen siebziger Jahren." *Geschichte im Westen* 19 (2004): 230–43.

Nye, David. *Electrifying America: Social Meanings of a New Technology, 1880–1949*. Cambridge, MA: MIT Press, 1990.

O'Dea, William T. *The Social History of Lighting*. London: Routledge and Kegan Paul, 1958.

Ó'Gráda, Cormac. *A Rocky Road: The Irish Economy since the 1920s*. Manchester, UK: Manchester University Press, 1997.

Oldenziel, Ruth. "Man the Maker, Woman the Consumer: The Consumption Junction Revisited," in *Feminism in Twentieth-Century Science, Technology, and Medicine*, edited by Angela N.H. Creager, Elizabeth Lunbeck, and Londa Schiebinger, 128–48. Chicago, IL: Chicago University Press, 2001.

Orrinsmith, Mrs [Lucy Faulkner]. *The Drawing-Room: Its Decorations and Furniture*. London: Macmillan and Company, 1878.

Ostwald, Wilhelm. "The Modern Theory of Energetics." *Monist* 17 (1907): 510–11.

Otter, Chris. *The Victorian Eye*. Chicago, IL: Chicago University Press, 2008.

Otter, Chris, Alison Bashford, John L. Brooke, Fredrik A. Jonsson, and Jason M. Kelly. "Roundtable: The Anthropocene in British History." *Journal of British Studies* 57, no. 3 (2018): 568–96.

Panton, J.E. [Jane Ellen]. *Suburban Residences and How to Circumvent Them*. Cambridge, UK: Cambridge University Press, (1896) 2012.

Parker, Rozsika, and Griselda Pollock. *Old Mistresses: Women, Art and Ideology*. London: Routledge and Kegan Paul, 1981.

Parr, Joy. *Domestic Goods: The Material, the Moral and the Economic in the Postwar Years*. Toronto, ON: University of Toronto Press, 1999.

– "Modern Kitchen, Good Home, Strong Nation." *Technology and Culture* 43, no. 4 (2002): 657–67.

– "Shopping for a Good Stove: A Parable about Gender, Design and the Market." In *A Diversity of Women: Women in Ontario since 1945*, edited by Joy Parr, 75–97. Toronto, ON: University of Toronto Press, 1995. Revised and reprinted in *His and Hers: Gender, Technology and Consumption*, edited by Roger Horowitz and Arwen Mohun, 165–88. Charlottesville, VA: University of Virginia Press, 1998.

– "What Makes Washday Less Blue? Gender, Nation, and Technology Choice in Postwar Canada." *Technology and Culture* 38, no. 1 (1997): 153–86.

Parsons, Katharine. "Women's Work in Engineering and Shipbuilding during the War." *Transactions of the North East Coast Institution of Engineers and Shipbuilder* 35 (1918–1919): 227–36.

Partridge, Margaret. "The Direct Current Dynamo, Notes on Its Construction and Habits." *Woman Engineer* 1 (1919–1920): 26–7.

Pateman, Carol. *The Disorder of Women: Democracy, Feminism and Political Theory*. Stanford, CA: Stanford University Press, 1989.

Payne, Peter L. *The Hydro: A Study of the Development of the Major Hydro-Electric Schemes undertaken by the North of Scotland Hydro-Electric Board*. Aberdeen, UK: Aberdeen University Press, 1988.

Pennington, Shelley, and Belinda Westover. *A Hidden Workforce: Homeworkers in England, 1850–1985*. London: Macmillan, 1989.

Petrie, Edna S. "A Demonstrator's Work in the North of Scotland." *Electrical Age* (1963): 496–8.

Petrocultures Research Group. *After Oil*. Edmonton, AB: University of Alberta, 2016.

Petzina, Dietmar. "Wirtschaft und Arbeit im Ruhrgebiet." In *Das Ruhrgebiet im Industriezeitalter: Geschichte und Entwicklung*, edited by Wolfgang Köllmann, et al., 491–567. Düsseldorf, Germany: Schwann, 1990.

Pinchbeck, Ivy. *Women Workers and the Industrial Revolution, 1750–1850.* London: Routledge, 1930.

Plotnick, Rachel. "At the Interface: The Case of the Electric Push Button, 1880–1923," *Technology and Culture* 53, no. 4 (2012): 815–45.

Pollard, Eliza F. *Florence Nightingale: The Wounded Soldier's Friend.* London: Partridge, 1895.

Pomeranz, Kenneth. *The Great Divergence: China, Europe and the Making of the Modern World Economy,* Princeton, NJ: Princeton University Press, 2000.

Portelli, Alessandro. "What Makes Oral History Different." In *The Oral History Reader,* edited by Robert Perks and Alistair Thomson. Abingdon, UK: Routledge, 2016, 48–58.

Pratt, Brenda M. "Home Economics Subject Development in the Context of Secondary Education." Unpublished MPhil thesis, University of Surrey, UK, 1990. Accessed 10 November 2019. http://epubs.surrey.ac.uk/859/1/fulltext.pdf.

Pursell, Carroll. "Domesticating Modernity: The Electrical Association for Women, 1924–86." *British Journal for the History of Science* 32, no. 1 (1999): 47–67.

Rathbone, Eleanor. *The Disinherited Family: A Plea for Direct Provision for the Costs of Child Maintenance through Family Allowances.* London: Gage Allen and Unwin, 1924.

Reddy, William. *The Rise of Market Culture: The Textile Trade in French Society, 1750–1900.* Cambridge, UK: Cambridge University Press, 1984.

Reeves, Maud Pember. *Round about a Pound a Week.* London: Virago Press, (1913) 1979.

REO News. "The Spring Show." *REO News,* 1948, 3.

Reynolds, Terry S. *Stronger than A Hundred Men: A History of the Vertical Water Wheel.* Baltimore, MD: Johns Hopkins University Press, 1983.

Richman Keneally, Rhona. "Tastes of Home in Mid-Twentieth-Century Ireland: Food, Design, and the Refrigerator." *Food and Foodways* 23 (2015): 1–24.

Roberts, Gerrylynn. "Electrification." In *Science, Technology and Everyday Life, 1870–1950,* edited by Colin Chant, 68–112. London: Routledge and Open University, 1989.

Robins, Jonathan. "Oil Boom: Agriculture, Chemistry, and the Rise of Global Plant Fat Industries, ca. 1850–1920." *Journal of World History* 3 (2018): 313.

Romines, Ann. *The Home Plot: Women, Writing and Domestic Ritual.* Amherst, MA: University of Massachusetts Press, 2014.

Rose, Sonya. *Limited Livelihoods: Gender and Class in Nineteenth Century England.* Berkeley, CA: University of California Press, 1992.

Rosen, Christine Meisner. "Knowing Industrial Pollution: Nuisance Law and the Power of Tradition in a Time of Rapid Economic Change, 1840–1864." *Environmental History* 8, no. 4 (2003): 565–97.

Ross, Ellen. *Love and Toil: Motherhood in Outcast London, 1870–1918*. New York: Oxford University Press, 1993.

Rubio, Maria del Mar, and Mauricio Folchi. "Will Small Energy Consumers Be Faster in Transition? Evidence from the Early Shift from Coal to Oil in Latin America." *Energy Policy* 50 (2012): 50–61.

Saelens, Wout. "Review of *The Path to Sustained Growth: England's Transition from an Organic Economy to an Industrial Revolution* (Edward Anthony Wrigley, 2016)." *Journal of Energy History/Revue d'histoire de l'énergie* 2 (2019).

Samuel, Raphael. "Workshop of the World: Steam Power and Hand Technology in Mid-Victorian Britain." *History Workshop Journal* 3 (1977): 6–72.

Sandwell, R.W. "The Coal-Oil Lamp." *Agricultural History* 92, no. 2 (spring 2018): 190–209.

– "The Emergence of Modern Lighting in Canada: A Preliminary Reconnaissance." *Extractive Industries and Society: An International Journal* 3, no. 3 (2016): 850–63.

– "How Households Shape Energy Transitions: Canada's Great Transformation." In "Energizing the Spaces of Everyday Life: Learning from the Past for a Sustainable Future," edited by Vanessa Taylor and Heather Chappells, special edition of RCC *Perspectives* 2 (2019): 23–30.

– "An Introduction to Canada's Energy History." In *Powering Up Canada: A History of Power, Fuel, and Energy from 1600*, edited by R.W. Sandwell, 3–36. Montreal and Kingston: McGill-Queen's University Press, 2016.

– "The Limits of Liberalism: The Liberal Reconnaissance and the History of the Family in Canada." *Canadian Historical Review* 84, no. 3 (2003): 423–50.

– "Pedagogies of the Unimpressed: Re-Educating Ontario Women for the Mineral Economy, 1900–1940." *Ontario History* 107, no. 1 (2015): 36–59.

– "People, Place and Power: Rural Electrification in Canada, 1890–1950." In *Transforming the Countryside: The Electrification of Rural Britain*, edited by Paul Brassley, Jeremy Burchardt, and Karen Sayer, 178–204. London and New York: Routledge. 2017.

Sarti, Raffaella, ed. "Men at Home: Domesticities, Authority, Emotions and Work." Special issue of *Gender and History* 27, no. 3 (2015): 521–886.

Savage, W.G. *Rural Housing*. London: T. Fisher Unwin, 1915.

Sayer, Karen. "'His Footmarks on Her Shoulders': The Place of Women within Poultry Keeping in the British Countryside, c.1880 to c.1980." *Agricultural History Review* 61, no. 2 (2013): 301–29.

– "Atkinson Grimshaw, *Reflections on the Thames* (1880): Explorations in the Cultural History of Light and Illumination." *Annali di Ca' Foscari. Serie Occidentale* 51 (2017).

Scharff, Virginia J. "Introduction." In *Seeing Nature Through Gender*, edited by Virginia J. Scharff. Lawrence, KS: University Press of Kansas, 2003.

Schivelbusch, Wolfgang. *Disenchanted Night: The Industrialisation of Light in the Nineteenth Century*. Oxford, UK: Berg, 1988.

Schneider, Sigrid, ed. *Als der Himmel blau wurde: Bilder aus den 60er Jahren/ Ruhrlandmuseum Essen*. Bottrop/Essen, Germany: Verlag Peter Pomp and Ruhrlandmuseum Essen, 1998.

– "Bild – Geschichte: Fotografien aus dem Ruhrgebiet der sechziger Jahre." In *Als der Himmel blau wurde: Bilder aus den 60er Jahren/Ruhrlandmuseum Essen*, edited by Sigrid Schneider, 11–18. Bottrop/Essen, Germany: Verlag Peter Pomp and Ruhrlandmuseum Essen, 1998.

Schöttler, Peter. "Die Geschichtswerkstatt e.V.: Zu einem Versuch, basisdemokratische Geschichtsinitiativen und -forschungen zu 'vernetzen.'" *Geschichte und Gesellschaft* 10, no. 3 (1984): 421–4.

Schuchmann, Rolf. "Kindheit in einer Bergmannsfamilie in den 60er Jahren." *Zeit-Räume Ruhr*. Accessed 8 July 2020. http://www.zeit-raeume.ruhr/ die-erinnerungsorte/bergarbeitersiedlung-dortmund-marten-hangeney- kindheit-in-einer-bergmannsfamilie-in-den-60er-jahren/.

Schwartz Cowan, Ruth. "The Consumption Junction: A Proposal for Research Strategies in the Sociology of Technology." In *The Social Construction of Technological Systems: New Directions in the Sociology and History of Technology*, edited by Wiebe E. Bijker, Thomas P. Hughes, and Trevor J. Pinch, 261–80. Cambridge, MA: MIT Press, 1987.

– *More Work for Mother: The Ironies of Household Technologies from the Open Hearth to the Mircowave*. New York: Basic Books, 1983.

Scott, Dayna Nadine, ed. *Our Chemical Selves: Gender, Toxics, and Environmental Health*. Vancouver, BC: University of British Columbia Press, 2015.

Scott, Joan. "Gender: A Useful Category of Historical Analysis." *American Historical Review* 91 (1986): 10.

– *Gender and the Politics of History*. New York: Columbia University Press, 1999.

Scott, Peggy. *An Electrical Adventure*. London: Electrical Association for Women, 1934.

Scottish Government, Housing and Social Justice Directorate. "Scottish House Condition Survey: 2017 Key Findings," 4 December 2018. Accessed 10 November 2019. https://www.gov.scot/publications/scottish-house- condition-survey-2017-key-findings/pages/6.

Sharpe, Pamela. *Adapting to Capitalism: Working Women in the English Economy, 1700–1850*. Basingstoke, UK: Macmillan, 1996.

Shelley, Mary. *Frankenstein, or the Modern Prometheus*. London: Lackington, Hughes, Harding, Mavor, and Jones, 1818.

Sherer, John. "The Management of the Home." *Family Friend* 3, no. 3 (1867): 201–3.

Shiel, Michael. *The Quiet Revolution: The Electrification of Rural Ireland.* Dublin: O'Brien Press, 1984.

Shove, Elizabeth, and Gordon Walker. "What Is Energy For? Social Practice and Energy Demand." *Theory, Culture and Society* 31 (2014): 41–58.

Shutt, Frank T. "The Air of Our Houses." *Ottawa Naturalist* 7, no. 2 (1893): 24–32.

Sieferle, Rolf Peter. *The Subterranean Forest: Energy Systems and the Industrial Revolution.* Translated by Michael P. Osman. Cambridge, UK: White Horse Press, 1982.

Simmons, Ian Gordon. *Environmental History: A Concise Introduction.* Oxford, UK: Blackwell, 1993.

Simon, Linda. *Dark Light: Electricity and Anxiety from the Telegraph to the X-Ray.* Orlando, FL: Harcourt, 2004.

Skeggs, Beverley. *Formations of Class and Gender: Becoming Respectable.* London: Sage, 1997.

Smil, Vaclav. "Energy Flows in the Developing World." *American Scientist* 67 (1979): 522–3.

Smith, Bonnie. "Havens No More: Discourses on Domesticity." *Gender and History* 2, no. 1 (spring 1990): 98–102.

Smith, Drummond. "The Housing of the Scottish Farm Servant." *Economic Journal* 25, no. 99 (1915): 466–74.

Smith, Mary E.H. *A Guide to Housing.* London: Housing Centre Trust, 1971.

Smout, T.C. *Nature Contested: Environmental History in Scotland and Northern England since 1600.* Edinburgh, UK: Edinburgh University Press, 2000.

SPAB Blog. "Women in Conservation: House of Garrett." 10 March 2016. https://thespab.wordpress.com/2016/03/10/women-in-conservation-house-of-garrett.

Spiegelberg, Friedrich. *Energiemarkt im Wandel: Zehn Jahre Kohlenkrise an der Ruhr.* Baden-Baden, Germany: Nomos, 1970.

Sprenger, Elizabeth, and Pauline Webb. "Persuading the Housewife to Use Electricity? An Interpretation of Material in the Electricity Council Archives." *British Journal for the History of Science* 26, no. 1 (1993): 55–65.

Spring Rice, Margery. *Working-Class Wives: Their Health and Conditions,* 2nd ed. London: Virago, (1939) 1981.

Stadt Recklinghausen. *Hochlarmarker Lesebuch, Kohle war nicht alles: 100 Jahre Ruhrgebietsgeschichte.* Oberhausen, Germany: Asso Verlag, 1981.

Stage, Sarah, and Virginia Vincenti, eds. *Rethinking Home Economics: Women and the History of a Profession.* Ithaca, NY: Cornell University Press, 1997.

Steinberg, Ted. *Nature Incorporated: Industrialization and the Waters of New England.* Cambridge, UK: Cambridge University Press, 1991.

Stephenson, Janet. 2018. "Sustainability Cultures and Energy Research: An Actor-Centred Interpretation of Cultural Theory." *Energy Research and Social Science* 44 (2018): 242–9.

Stewart, Mary Lynn. *Women, Work and the French State: Labour Protection and Social Patriarchy 1879–1919*. Montreal and Kingston: McGill Queen's University Press, 1989.

Strasser, Susan. *Never Done: A History of American Housework*. New York: Pantheon Books, 1982.

Strong-Boag, Veronica. *The New Day Recalled: Lives of Girls and Women in English Canada, 1919–1939*. Toronto, ON: Copp, Clark, Pitman, 1993.

Sunlight Year Book. Port Sunlight, UK: Lever Brothers, 1898.

Szeman, Imre, Jennifer Wenzel, and Patricia Yaeger, eds. *Fuelling Culture: 101 Words for Energy and Environment*. New York: Fordham University Press, 2017.

Tanner, Ariane. "Thinking with Energy: Holism and the History of Energetics." Forum: Environmental History of Energy Transitions, *Environmental History* 24, no. 3, (2019): 482–91.

Tarr, Joel. "Housewives as Home Safety Managers: The Changing Perception of the Home as a Place of Hazard and Risk, 1870–1940." In *Accidents in History*, edited by Roger Cooter and Bill Luckin, 196–233. Amersterdam: Rodopi, 1995.

Taylor, Vanessa, and Heather Chappells. "What Consumers in the Past Tell Us about Future Energyscapes." In "Energizing the Spaces of Everyday Life: Learning from the Past for a Sustainable Future," edited by Vanessa Taylor and Heather Chappells, special edition of RCC *Perspectives* 2 (2019): 11–21.

Tegetmeier, William Bernard. *A Manual of Domestic Economy with Hints on Domestic Medicine and Surgery*. London: Home and Colonial School Society, 1880.

Tenfelde, Klaus. "Vom Ende und Anfang sozialer Ungleichheit: Das Ruhrgebiet in der Nachkriegszeit," special issue of *Geschichte und Gesellschaft* 22 (2006): 269–85.

Thompson, E.P. *The Making of the English Working Class*. New York: Pantheon, 1963.

– *Whigs and Hunters: The Origin of the Black Act*. New York: Pantheon, 1975.

Thompson, Paul. "The Voice of the Past: Oral History." In *The Oral History Reader*, edited by Robert Perks and Alistair Thomson, 33–9. Abingdon, UK: Routledge, 2016.

Thompson, Phyllis. "Women on Electricity Committees." *Electrical Age* (winter/spring 1938): 334–5.

Thomson, J.H., and Boverton Redwood. *The Petroleum Lamp, Its Choice and Use: A Guide to the Safe Employment of Mineral Oil in What Is Commonly Termed the Paraffin Lamp*. London: Charles Griffin and Co. Exeter St. Strand, 1902.

Thresh, John C. *The Housing of the Agricultural Labourer, with Special Reference to Essex*. London: Rural Housing and Sanitation Association, 1919.

Tilly, Louise A., and Joan W. Scott. *Women, Work and Family*. New York and London: Holt, Rinehart and Winston, 1978.

Tivey, Leonard. "Quasi-Government for Consumers." In *Quangos in Britain: Government and the Networks of Public Policy-Making*, edited by Anthony Barker, 137–51. London: Macmillan, 1982.

Tobey, Ronald C. *Technology as Freedom: The New Deal and the Electrical Modernization of the American Home*. Berkeley, CA: University of California Press, 1996.

Tomes, Nancy. *The Gospel of Germs: Men, Women and the Microbe in American Life*. Cambridge, MA: Harvard University Press, 1998.

– "The Private Side of Public Health: Sanitary Science, Domestic Hygiene and the Germ Theory, 1870–1900." *Bulletin of the History of Medicine* 64 (winter 1990): 509–39.

Toynbee, Arnold. *The Industrial Revolution*, Boston, MA: Beacon Press, 1960.

Trentmann, Frank. "Getting to Grips with Energy: Fuel, Materiality and Daily Life." In "The Material Culture of Energy," special issue of *Science Museum Group Journal* (spring 2018): 8.

– ed. *The Making of the Consumer: Knowledge, Power and Identity in the Modern World*. Oxford and New York: Berg, 2006.

Trentmann, Frank, and Anna Carlsson-Hyslop. "The Evolution of Energy Demand in Britain: Politics, Daily Life, and Public Housing, 1920s–1970s." *Historical Journal* 61, no. 3 (2018): 807–39.

Unger, Nancy. "Women and Gender: Useful Categories of Analysis in Environmental History." *Oxford Handbook of Environmental History*, edited by Andrew C. Isenberg (unpaginated). Oxford, UK: Oxford University Press, 2014.

Unger, Richard W., and John Thistle. *Energy Consumption in Canada in the Nineteenth and Twentieth Centuries: A Statistical Outline*. Rome: CNR Edizioni, 2013.

Valencius, Conevery Bolton. *The Health of the Country: How American Settlers Understood Themselves and Their Land*. New York: Basic Books, 2002.

Valenze, Deborah. *The First Industrial Woman*. Oxford, UK: Oxford University Press, 1995.

Verbeeck, Peter-Paul. *What Things Do: Philosophical Reflections on Technology, Agency and Design*. University Park, PA: Pennsylvania State University Press, 2005.

Von Plato, Alexander, ed. *"Der Verlierer geht nicht leer aus": Betriebsräte geben zu Protokoll*. Bonn, Germany: Dietz, 1984.

Waldram, James B. *As Long as the Rivers Run: Hydroelectric Development and Native Communities in Western Canada*. Winnipeg, MB: University of Manitoba Press, 1988.

Walker, Lynne. "The Arts and Crafts Alternative." In *A View from the Interior: Women and Design*, edited by Judy Attfield and Pat Kirkham, 163–74. London: Women's Press, 1989.

– "Women Architects." In *A View from the Interior: Women and Design*, edited by Judy Attfield and Pat Kirkham, 90–108. London: Women's Press, 1989.

Walsh, Margaret, and Chris Wrigley. "Womanpower: The Transformation of the Labour Force in the UK and the USA since 1945." *REFRESH* 30 (summer 2001): 1–4.

Warde, Paul. "The Hornmoldt Metabolism: Energy, Capital and Time in an Early Modern German Household." Forum: Environmental History of Energy Transitions, *Environmental History* 24, no. 3 (2019): 472–81.

White, Leslie A. "Energy and the Evolution of Culture." *American Anthropologist* 45 (July–September 1943): 335–56.

– *The Science of Culture: A Study of Man and Civilization*. New York: Grove Press, 1949.

White, Richard. "Dying for Progress." In *The Republic for Which It Stands: The United States during Reconstruction and the Gilded Age, 1865–1896*, 477–517. New York: W.W. Norton, 2017.

– *The Organic Machine: The Remaking of the Columbia River*. New York: Hill and Wang, 1996.

– *The Republic for Which It Stands: The United States during Reconstruction and the Gilded Age, 1865–1896*. New York: W.W. Norton, 2017.

Whittle, Jane. "A Critique of Approaches to 'Domestic Work': Women, Work and the Pre-Industrial Economy." *Past and Present*, no. 243 (2019): 35–70.

Wierling, Dorothee. "Alltagsgeschichte und Geschichte der Geschlechterbeziehungen: Über historische und historiographische Verhältnisse." In *Alltagsgeschichte: Zur Rekonstruktion historischer Erfahrungen und Lebensweisen*, edited by Alf Lüdtke, 169–90. Frankfurt, Germany: Campus Verlag, 1989.

Wilhite, Harold. "Energy Consumption As Cultural Practice: Implications for the Theory and Policy of Sustainable Energy Use." In *Cultures of Energy: Power, Practices, Technologies*, edited by Sarah Strauss, Stephanie Rupp, and Thomas Love, 60–72. Walnut Creek, CA: Left Coast Press, 2013.

Williams, James. *Energy and the Making of Modern California*. Akron, OH: University of Akron Press, 1997.

Williamson, Harold R., and Arnold R. Daum. *The American Petroleum Industry: Volume 1: The Age of Illumination*. Evanston, IL. Northwestern University Press, 1959.

Wlasiuk, Jonathan Joseph. *Refining Nature: Standard Oil and the Limits of Efficiency*. Pittsburgh, PA: University of Pittsburgh Press, 2017.

Wood, Ellen Meiksins. *The Origin of Capitalism: A Longer View*. London and New York: Verso, 2017.

Worden, Suzette. "Powerful Women: Electricity in the Home, 1919–40." In *A View From the Interior: Feminism, Women and Design*, edited by Judy Attfield and Pat Kirkham, 131–40. London: Women's Press, 1989.

Wrigley, E.A. *Continuity, Chance and Change: The Character of the Industrial Revolution in England.* Cambridge, UK: Cambridge University Press, 1988.

– *Energy and the English Industrial Revolution.* Cambridge, UK: Cambridge University Press, 2010.

– *The Path to Sustained Growth: England's Transition from an Organic Economy to an Industrial Revolution.* Cambridge, UK: Cambridge University Press, 2016.

– "The Supply of Raw Materials in the Industrial Revolution." *Economic History Review* 15 (1962): 1–16.

Wynn, Graeme. *Canada and Arctic North America, An Environmental History.* Santa Barbara, CA: ABC-CLIO, 2007.

Yong-Sook, Jung. "Just a Housewife? Miners' Wives between Household and Work in Postwar Germany." In *Mining Women*, edited by Jaclyn J. Gier and Laurie Mercier, 262–79. New York: Palgrave Macmillan, 2006.

Zachmann, Karin. "Introduction." In *Past and Present Energy Societies: How Energy Connects Politics*, edited by Nina Möllers and Karin Zachmann. Bielefeld, Germany: Transcript Verlag, 2012.

Zallen, Jeremy. *American Lucifers: The Dark History of Artificial Light.* Chapel Hill, NC: University of North Carolina Press, 2019.

Zelizer, Viviana A. *Pricing the Priceless Child: The Changing Value of Children.* New York: Basic Books, 1985.

Zimring, Carl A. *Aluminum Upcycled: Sustainable Design in Historical Perspective.* Baltimore, MD: Johns Hopkins University Press, 2017.

Contributors

PETRA DOLATA is an associate professor of history at the University of Calgary and scholar in residence at the Calgary Institute for the Humanities, where she co-convenes the Energy in Society working group. She is the author of *Die deutsche Kohlenkrise im nationalen und transatlantischen Kontext* (The German Coal Crisis and Its National and Transatlantic Dimensions), published in 2006. Her research focuses on European and North American energy history after 1945, specifically the history of energy transitions and the 1970s energy crises.

GRAEME GOODAY is a professor of history of science and technology in the School of Philosophy, Religion and History of Science at the University of Leeds. His main research has been on the history of electrical science and technology in Britain between the 1870s and 1920s, focusing on issues of trust, expertise, gender, and patents, recently developing an interest in the historical relationship between hearing loss, acoustics, and hearing aids. His publications include *The Morals of Measurement: Accuracy, Irony, and Trust in Late Victorian Electrical Practice* (Cambridge University Press, 2004); *Domesticating Electricity: Technology, Uncertainty and Gender, 1880–1914* (2008, republished in paperback by University of Pittsburgh Press, 2018); *Patently Contestable: Electrical Technologies and Inventor Identities on Trial in Britain* (MIT Press, 2013), coauthored with Stathis Arapostathis; and *Managing the Experience of Hearing Loss in Britain, 1830–1930* (Palgrave Macmillan, 2017), coauthored with Karen Sayer. He collaborates extensively on public engagement with museums, and in 2019 worked with the UK Women's Engineering Society in commemorating its centenary in the "Electrifying Women" project.

ABIGAIL HARRISON MOORE is a professor of art history and museum
studies at the University of Leeds, where she was the head of the School
of Fine Art, History of Art and Cultural Studies until August 2019.
Previous publications include *Fraud, Fakery and False Business: Rethinking
the Shrager v. Dighton "Old Furniture Case"* (Continuum, 2011) and, with
Dorothy Rowe, *Architecture and Design in Europe and America, 1550–2000*
(Blackwells, 2006). In terms of energy studies, she has published a
number of articles on Arts and Crafts approaches to designing for
energy transitions in the nineteenth century, and coedited with Ruth
Sandwell, *Off-Grid Empire: Rural Energy Consumption in Britain and the
British Empire, 1850–1960* (special issue of the *History of Retailing and
Consumption* 4, no. 1, 2018). She is currently working on a book on the
impact of women decorators on energy decisions, provisionally titled
*Switching from Master to Mistress: Women as Advisors and Consumers in
Energy Decisions, 1870–1910.*

SORCHA O'BRIEN is a design historian interested in technology and
identity, in both physical and digital forms, who teaches design history
at the National College of Art and Design in Dublin, Ireland. Prior to
this, she was a senior lecturer in design history and theory at Kingston
University, London. She was an AHRC Early Career Leadership Fellow
from 2016 to 2019, working on the introduction of electrical products
into the Irish home in the wake of rural electrification in the 1950s and
1960s, in partnership with the National Museum of Ireland. Outputs
from this project include the *Kitchen Power: Women's Experiences of Rural
Electrification* exhibition at the National Museum of Ireland–Country
Life and a forthcoming accompanying monograph. She is the author of
Powering the Nation: Images of the Shannon Scheme and Electricity in Ireland
(Irish Academic Press, 2017) and coeditor of *Love Objects: Emotion, Design
and Material Culture* (Bloombury Academic, 2013) and the *Bloomsbury
Encyclopedia of Design* (Bloomsbury Academic, 2015).

RUTH W. SANDWELL is a historian and history educator at the University
of Toronto. In 2019 she was a fellow at the Rachel Carson Center for
Environment and Society. She is the author of a number of articles
exploring the history of energy and everyday life in Canada in the nineteenth and twentieth centuries. She was coeditor with Abigail Harrison
Moore of *Off-Grid Empire: Rural Energy Consumption in Britain and the
British Empire, 1850–1960* (special issue of the *History of Retailing and
Consumption* 4, no. 1, 2018), and edited *Powering Up Canada: A History
of Power, Fuel, and Energy from 1600* (McGill-Queen's University Press,
2016). She authored the monograph *Canada's Rural Majority: Households,*

Environments, Economies, 1870–1940 (University of Toronto Press, 2016), and is currently finishing up *Heat, Light and Work in the Home: A Social History of Energy, 1800–1940*.

KAREN SAYER is a professor of social and cultural history at Leeds Trinity University, UK. She is also codirector of the Leeds Centre for Victorian Studies. Within the Leeds Centre for Victorian Studies and its wider networks, she draws on material culture, illustration, and text to work on Victorian social and cultural histories of landscapes of marginal spaces and experiences, such as nocturnal landscapes of waterways, rivers, and coastlines, material technologies of sight and sound, cultures of light and illumination, and the aesthetics and material cultures of hearing loss. The key interlinking theme of her research is the ways in which bodies, materials, and environments are shaped in the nineteenth and twentieth centuries.

VANESSA TAYLOR is a historian based at the University of Greenwich in London, UK. She has most recently published on energy, in the RCC *Perspectives* issue (2020, no. 1) that paved the way for this volume, and on water and its meanings in *A Mighty Capital under Threat: The Environmental History of London, 1800–2000*, edited by Bill Luckin and Peter Thorsheim (University of Pittsburgh Press, 2020). She is currently writing a book on rivers.

Index

copper, as light reflector, 47, 103
cotton mills, 28, 56; cotton as an oil,
    47
COVID-19, 43n62, 53n74, 67, 112n57
craft, 10, 91, 104, 116
Cragside, 103. *See also* Armstrong,
    Sir William
creosote, 73
Crompton, 107

danger: danger with home
    appliances, 67–89; danger of
    dishonest design, 104; industrial
    danger, 32; and new energy forms,
    8, 12, 27–8, 48, 67–89. *See also* fear
Davidson, Caroline, *A Woman's Work
    Is Never Done*, 30, 33
daylight, 46, 50, 98
death: death rates, 28, 72–3, 79; as a
    result of accidents in energy use,
    67–89; women's death due to
    overwork, 167
decorators, decorating, 90–113
deficit accounts, in understanding
    energy transitions, 13, 93, 117
de Jong, Jutta, 160
demonstrators, 5, 174, 176; of electric
    appliances, 115, 128, 141, 152n22;
    of gas appliances, 76–7, 126
designers, 6, 9, 90–113
design history, 134, 138
Diploma for Demonstrators and
    Saleswomen, 128
Diploma for Teaching Electrical
    Housecraft, 128
dirt: as energy byproduct, 76, 79, 101,
    161, 165, 167, 181; as pollution,
    155, 167. *See also* cleaning:
    cleanliness; pollution, air
disease, 32, 81; miasmic theories
    of, 69, 77–8, 181. *See also* air;
    carbonic acid; health

dishwashers. *See* appliances
district health visitors, 77. *See also*
    health
Domestic Coal Consumers' Council,
    183
domestic science, 5. *See also* home
    economics
domestic work. *See* drudgery;
    housework; servants
Digital Repository of Ireland Life
    Histories and Social Change
    Collection, 139
drudgery: of housework, 32, 116–17,
    124, 130, 143–4, 148–50. *See also*
    housewives; housework
dryers. *See* appliances
Durken, Noreen, 134, 142

EAW certificate examination, 128
Edis, Robert, *Our Homes and How to
    Make them Healthy*, 104
Edwards, Elsie, 129
Electrical Age, The, 176
electrical appliances. *See* appliances
Electrical Association for Women
    (EAW), 114–33, 174, 176, 182–4.
    See also *Electrical Handbook for
    Women*; Haslett, Caroline
Electrical Development Association,
    128, 135
Electrical Handbook for Women,
    114–17, 122, 125–6, 128–9
Electrical Lamp Manufacturers
    Association, 128
Electric Irish Homes Project, 138–9,
    152n21
electricity, 21–2; British national
    grid, 5; design for electricity,
    90–107; electricial engineering,
    10, 114–33; electricity bills,
    142–3; electricity consumption
    and supply, 4–5, 7, 10–11,